热　　学

Thermal Physics

张博群　夏蔡娟　编著

西北工业大学出版社

西　安

【内容简介】 本书主要介绍分子电子学中的热学和热传输特性。内容包括温度、简单热力学系统、热力学关系、开放系统、临界现象、化学平衡和异构系统等内容。

本书可作为高等院校物理学、应用物理学以及工程热物理相关专业的参考书。

图书在版编目(CIP)数据

热学=Thermal Physics / 张博群,夏蔡娟编著
.—西安:西北工业大学出版社, 2020.6
ISBN 978-7-5612-6813-1

Ⅰ.①热… Ⅱ.①张… ②夏… Ⅲ.①热学 Ⅳ.
①O551

中国版本图书馆CIP数据核字(2020)第067244号

RE XUE
热 学

责任编辑: 朱辰浩	**策划编辑:** 季 强
责任校对: 张 潼	**装帧设计:** 李 飞
出版发行: 西北工业大学出版社	
通信地址: 西安市友谊西路127号	邮编:710072
电 话: (029)88491757, 88493844	
网 址: www.nwpup.com	
印 刷 者: 陕西奇彩印务有限责任公司	
开 本: 787 mm×960 mm	1/16
印 张: 9.125	
字 数: 210千字	
版 次: 2020年6月第1版	2020年6月第1次印刷
定 价: 48.00元	

Preface

In this book, I have tried to do justice to both thermodynamics and statistical mechanics, without giving undue emphasis to either. The book is in three part. Part I including Chapter 1 and Chapter 2 introduces the fundamental principles of thermal physics in a unified way, going back and forth between the microscopic and macoscopic viewpoints. This portion of the book also applies these principles to a few simple thermodynamic systems, chosen for their illustrative character. Part Ⅱ and Ⅲ including Chapter 3 – Chapter 6 then develop more sophisticated techniques to treat further applications of thermodynamics and statistical mechanics, respectively. My hope is that this organizational plan will accomodate a variety of teaching philosophies in the middle of the thermo-to-statmech continuum. In structors who are entrenched at one or the other extreme should look for a different book.

One of my goals in writing this book was to keep it short enough for a one-semester course. Too many topics have made their way into the text, and it is now too long even for a very fast-paced semester. The book is still intended primarily for a one-semester course, however.

I was having great respect for the strong support of Xi'an Polytechnic University. Furthermore, I also acknowledge and express my deep sense of gratitude to individuals below for their valuable discussions and help: Professor Zhang Yingtang, Zhai Xuejun, Cheng Pengfei, Liu Hancheng, Zhang Guoqing, Li Lianbi, Su Yaoheng and Wang Jun.

I would greatly appreciate any comments and suggestions for improvements. Although extreme care was taken to correct all the misprints, it is very likely that I have missed some of them. I shall be most grateful to those readers who are kind enough to bring to my notice any remaining mistakes, typographical or otherwise for remedial. Please feel free to contact me.

Author

2020. 3

Contents

Chapter 1 Energy in Thermal Transport

Thermal transport is the fruit of the union of statistical and mechanical principles. Mechanics tells us the meaning of work; thermal physics tells us the meaning of heat. There are three new quantities in thermal physics that do not appear in ordinary mechanics: entropy, temperature, and free energy. We shall motivate their definitions in the first three chapters and deduce their consequences thereafter.

Our point of departure for the development of thermal physics is the concept of the stationary quantum states of a system of particles. When we can count the quantum states accessible to a system. we know the entropy of the system, for the entropy is defined as the logarithm of the number of states. The dependence of the entropy on the energy of the system defines the temperature. From the entropy. the temperature, and the free energy we find the pressure, the chemical potential, and all other thermodynamic properties of the system.

For a system in a stationary quantum state, all observable physical properties such as the energy and the number of particles are independent of the time. For brevity we usually omit the word stationary; the quantum states that we treat are stationary except when we discuss transport processes. The systems we discuss may be composed of a single particle or, more often, of many particles. The theory is developed to handle general systems of interacting particles, but powerful simplifications can be made in special problems for which the interactions may be neglected.

Each quantum state has a definite energy, States with identical energies are said to belong to the same energy level. The multiplicity or degeneracy of an energy level is the number of quantum states with very nearly the same energy. It is the number of quantum states that is important in thermal physics, not the number of energy levels. We shall frequently deal with sums over all quantum states. Two states at the same energy must always be counted as two states, not as one level.

Let us look at the quantum states and energy levels of several atomic systems. The simplest is hydrogen, with one electron and one proton. The low-lying energy level of hydrogen are shown in Figure 1. 1. The zero of energy in the figure is taken at the state of lowest energy. The number of quantum states belonging to the same energy level is in parentheses. In

Figure 1. 1 we overlook that the proton that the proton has a spin of 0. 5 h and has two independent orientations, parallel or antiparallel to the direction of an arbitrary external axis, such as the direction of a magnetic field. To take account of the two orientations we should double the values of the multiplicities shown for atomic hydrogen.

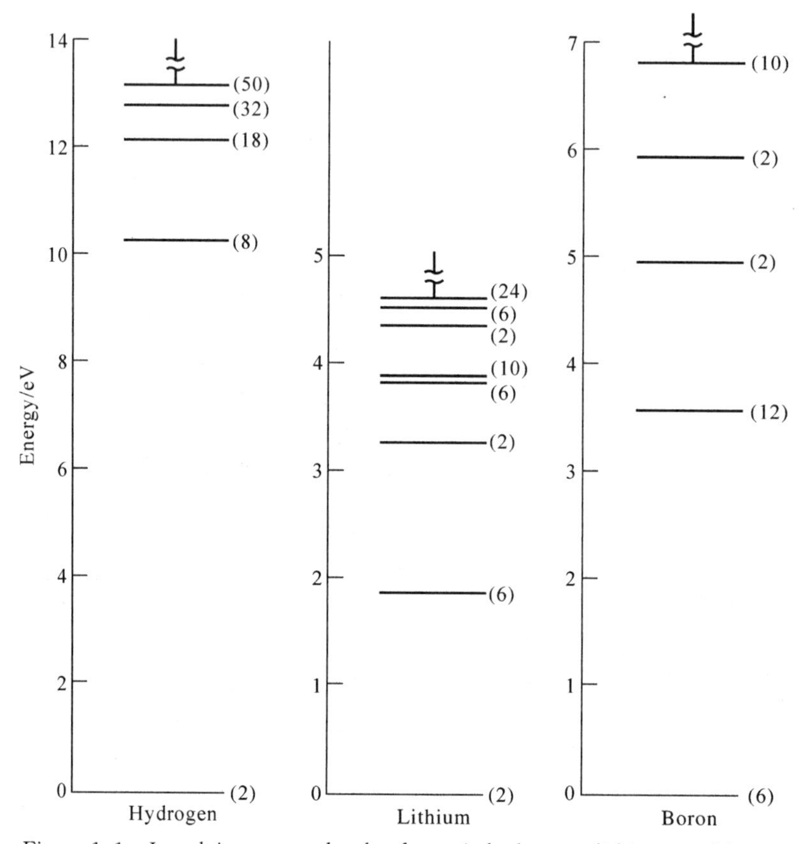

Figure 1. 1 Low-lying energy levels of atomic hydrogen, lithium, and boron

An atom of lithium has three electrons which move about the nucleus. Eachelectron interacts with the nucleus, and each electron also interacts with all the other electrons. The energies of the levels of lithium shown in Figure 1. 1 are the collective energies of the entire system. The energy levels shown for boron, which has five electrons are also the energies of the entire system.

The energy of a system is the total energy of all particles, kinetic plus potential, with account taken of interactions between particle. A quantum state of the system is asale of all particles Quantum states of a one-particle system are called orbitals. The low-lying energy levels of a single particle of mass M confined to a cube of side L are shown in Figure 1. 2. We shall find in Chapter 3 that an orbital of a free particle can be characterized by three positive integral quantum numbers n_x, n_y, n_z. The energy is

$$\varepsilon = \frac{\hbar^2}{2M}\left(\frac{\pi}{L}\right)^2 (n_x^2) + (n_y^2) + (n_z^2)$$

The multiplicities of the levels are indicated in Figure 1.2. The three orbitals with (n_x, n_y, n_z) equal to (4,1,1)(1,4,1) and (1,1,4) all have $n_x^2+n_y^2+n_z^2=18$; the corresponding energy level has the multiplicity 3.

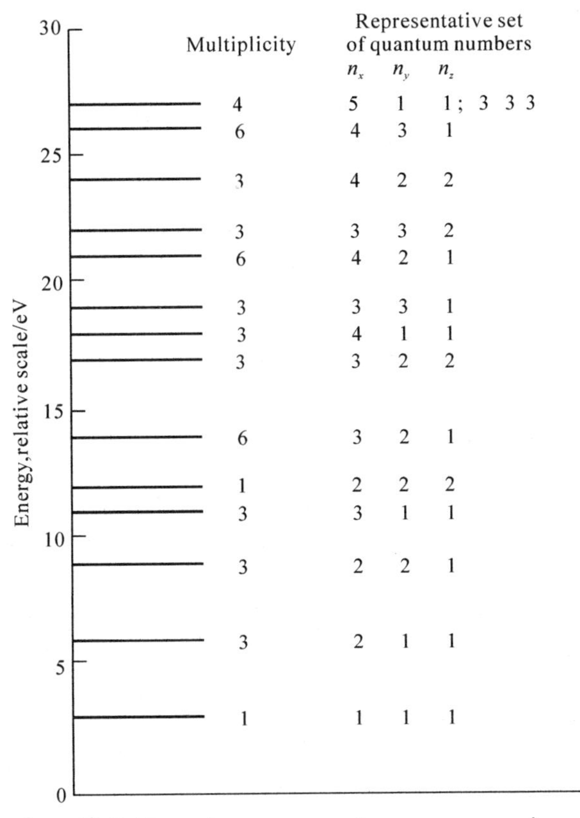

Figure 1.2 Energy levels multiplicities and quantum numbers n_x, n_y, n_z of a particle confined to a cube

To describe the statistical properties of a system of N particles, it is essential to know the set of values of the energy ,where is the energy of the quantum states of the N particle system. Indices such as s may be assigned to the quantum states in any convenient arbitrary way, but two different states should not be assigned the same index.

It is a good idea to start our program by studying the properties of simple model systems for which the energies $c_s(N)$ can be calculated exactly. We choose as a model a simple binary system because the general statistical properties found for the model system are believed to apply equally well to any realistic physical system. This assumption leads to predictions that always agree with experiment. What general statistical properties are of concern will become clear as we go along.

1.1 Binary Model Systems

The binary model system is illustrated in Figure 1.3. We assume there are N separate and distinct sites fixed in space, shown for convenience on a line. Attached to each site is an elementary magnet that can point only up or down, corresponding to magnetic moments m. To understand the system means to count the states. This requires no knowledge of magnetism; an element of the system can be any site capable of two states, labeled as yes or no, red or blue, occupied or unoccupied, zero or one, plus one or minus one. The sites are numbered, and sites with different numbers are supposed not to overlap in physical space. You might even think of the sites as numbered parking spaces in a car parking lot, as in Figure 1.4. Each parking space has two states, vacant or occupied by one car.

Whatever the nature of our objects, we may designate the two states by arrows that can only point straight up or straight down. If the magnet points up, we say that the magnetic moment is $+m$. If the magnet points down, the magnetic moment is $-m$.

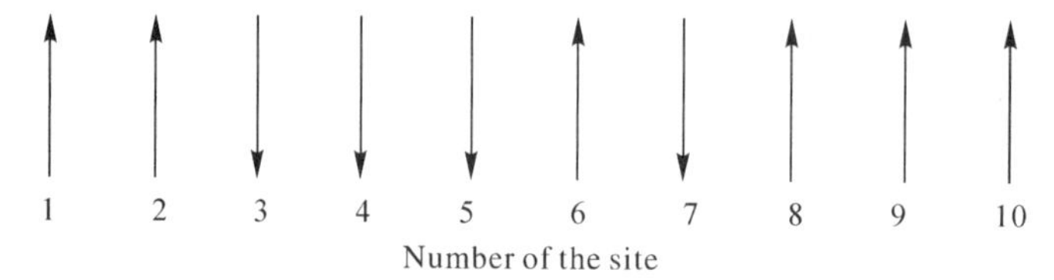

Figure 1.3 Model system composed of 10 elementary magnets at fixed sites on a line

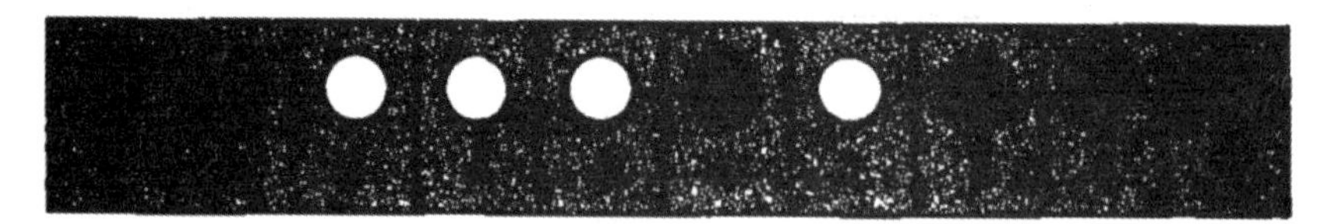

Figure 1.4 State of parking lot with 10 numbered parking spaces

Now consider N different sites, each of which bears a moment that may assume the values $\pm m$. Each moment may be oriented in two ways with a probability independent of the orientation of all other moments. The total number of arrangements of the N moments is $2\times 2\times\cdots\times 2=2^N$. A state of the system is specified by giving the orientation of the moment on each site; there are 2^N states.

$$\uparrow\uparrow\downarrow\downarrow\downarrow\uparrow\downarrow\uparrow\uparrow\uparrow \tag{1.1}$$

We may use the following simple notation for a single state of the system of N sites.

The sites themselves are assumed to be arranged in a definite order. We may number them in sequence from left to right, as we did in Figure 1. 3. According to this convention the state of Eq. (1. 1) also can be written as

$$\uparrow_1 \uparrow_2 \downarrow_3 \downarrow_4 \downarrow_5 \uparrow_6 \downarrow_7 \uparrow_8 \uparrow_9 \uparrow_{10} \cdots \tag{1. 2}$$

Both sets of symbols of Eq. (1. 1) and Eq. (1. 2) denote the same state of the system, the state in which the magnetic moment on site1 is $+m$; on site 2, the moment is $+m$; on site 3, the moment is $-m$; and so forth.

It is not hard to convince yourself that every distinct state of the system is contained in a symbolic product of N factors:

$$(\downarrow_1 + \downarrow_1)(\uparrow_2 + \downarrow_2)(\uparrow_3 + \downarrow_3)\cdots(\uparrow_N + \downarrow_N) \tag{1. 3}$$

The multiplication rule is defined by

$$(\uparrow_1 + \downarrow_1)(\uparrow_2 + \downarrow_2) = \uparrow_1 \uparrow_2 + \uparrow_1 \downarrow_2 + \downarrow_1 \uparrow_2 + \downarrow_1 \downarrow_2 \tag{1. 4}$$

The function of Eq. (1. 3) on multiplication generates a sum of 2^N terms. one for each of the 2^N possible states. Each term is a product of N individual magnetic moment symbols, with one symbol for each elementary magnet on the line. Each term denotes an independent state of the system and is a simple product of the form $\uparrow_1 \uparrow_2 \downarrow_3 \cdots \uparrow_N$, for example.

For a system of two elementary magnets, we multiply$(\uparrow_1 + \downarrow_1)by(\uparrow_2 + \downarrow_2)$to obtain the four possible states of Figure 1. 5.

$$(\uparrow_1 + \downarrow_1)(\uparrow_2 + \downarrow_2) = \uparrow_1 \uparrow_2 + \uparrow_1 \downarrow_2 + \downarrow_1 \uparrow_2 + \downarrow_1 \downarrow_2 \tag{1. 5}$$

1 2 1 2 1 2 1 2

Figure 1. 5 The four different states of a system of two elements numbered 1 and 2, where each element can have two conditions. The element is a magnet which can be in condition $\uparrow$ or condition $\downarrow$

The sum is not astate, but is a way of listing the four possible states of the system. The product on the left-hand side of the equation is called a generating function: it generates the states of the system.

The generating function for the states of a system of three magnets is

$$(\uparrow_1 + \downarrow_1)(\uparrow_2 + \downarrow_2)(\uparrow_3 + \downarrow_3)$$

This expression on multiplication generates $2^3 = 8$ different states:

Three magnets up: $\uparrow_1 \uparrow_2 \uparrow_3$

Two magnets up: $\uparrow_1 \uparrow_2 \downarrow_3 \quad \uparrow_1 \downarrow_2 \uparrow_3 \quad \downarrow_1 \uparrow_2 \uparrow_3$

One magnet up: $\uparrow_1 \downarrow_2 \downarrow_3 \quad \downarrow_1 \uparrow_2 \downarrow_3 \quad \downarrow_1 \downarrow_2 \uparrow_3$

None up: $\downarrow_1 \downarrow_2 \downarrow_3$

The total magnetic moment of our model system of N magnets each of magnetic moment m will be denoted by M, which we will relate to the energy in a magnetic field. The value of M varies from Nm to $-Nm$. The set of possible values is given by

$$M = Nm, (N-2)m, (N-4)m, (N-6)m, \cdots, -Nm \tag{1.6}$$

The set of possible values of M is obtained if we start with the state for which all magnets are up ($M=-Nm$) and reverse one at a time. We may reverse N magnets to obtain the ultimate state for which all magnets are down ($M=-Nm$).

There are $N+1$ possible values of the total moment, whereas there are 2^N states. When $N \gg 1$, we have $2^N \gg N+1$. There are many more states than values of the total moment. If $N=10$, there are $2^{10}=1,024$ states distributed among 11 different values of the total magnetic moment. For large N many different states of the system may have the same value of the total moment M. We will calculate in the next section how many states have a given value of M.

Only one stale of a system has the moment $M=Nm$; that state is

$$\uparrow \uparrow \uparrow \uparrow \cdots \uparrow \uparrow \uparrow \uparrow \tag{1.7}$$

There are N ways to form a state with one magnet down:

$$\downarrow \uparrow \uparrow \uparrow \cdots \uparrow \uparrow \uparrow \uparrow \tag{1.8}$$

is one such stale; another is

$$\uparrow \downarrow \uparrow \uparrow \cdots \uparrow \uparrow \uparrow \uparrow \tag{1.9}$$

and the other states with one magnet down are formed from (1.7) by reversing any single magnet. The states of Eq. (1.8) and Eq. (1.9) have total moment $M=Nm-2m$.

1.2 Enumeration of States and the Multiplicity Function

We use the word spin as a shorthand for elementary magnet. It is convenient to assume that N is an even number, We need a mathematical expression for the number of states with $N_\uparrow = 1/2N+s$ magnets up and $N_\downarrow = 1/2N-s$ magnets down, where s is an integer. When we turn one magnet from the up to the down orientation, $1/2\ N+s$ goes to $1/2\ N+s-1$ and $1/2\ N-s$ goes to $1/2\ N-s+1$. The difference (number up$-$number down) changes from $2s$ to $2s-2$. The difference

$$N_\uparrow - N_\downarrow = 2s \tag{1.10}$$

is called the spin excess. The spin excess of the 4 states in Figure 1.5 is 2,0,0 - 2 from left to right. The factor of 2 in Eq. (1.10) appears to be a nuisance at this stage, but it will prove to be convenient.

The product in Eq. (1.3) may be written symbolically as

$$(\uparrow + \downarrow)^N$$

We may drop the site labels (the subscripts) from Eq. (1.3) when we are interested only in how many of the magnets in a sates are up and down, and not in which particular sites have magnets up and down. It we drop the labels and neglect the order in which the arrows appear in a given product, then Eq. (1.4) becomes

$$(\uparrow + \downarrow)^2 = \uparrow\uparrow + 2\uparrow\downarrow + \downarrow\downarrow$$

further,

$$(\uparrow + \downarrow)^3 = \uparrow\uparrow\uparrow + 3\uparrow\uparrow\downarrow + 3\uparrow\downarrow\downarrow + \downarrow\downarrow\downarrow$$

We find $(\uparrow + \downarrow)^N$ for arbitrary N by the binomial expansion

$$x + y^N = x^N + Nx^{N-1}y + \frac{1}{2}N(N-1)x^{N-2}y^2 + \cdots + y^N =$$

$$\sum_{t=0}^{N} \frac{N!}{(N-t)!} x^{N-t} y^t \tag{1.11}$$

We may write the exponents of x and y in a slightly different but equivalent form by replacing t with $\frac{1}{2}N - s$:

$$x + y^N = \sum_{s=-\frac{1}{2}N}^{\frac{1}{2}N} \frac{N!}{\left(\frac{1}{2}N + s\right)!\left(\frac{1}{2}N - s\right)!} x^{\frac{1}{2}N+s} y^{\frac{1}{2}N-s} \tag{1.12}$$

With this result the symbolic expression $(\uparrow + \downarrow)^N$ becomes

$$(\uparrow + \downarrow)^N \equiv \sum_s \frac{N!}{\left(\frac{1}{2}N + s\right)!\left(\frac{1}{2}N - s\right)!} \uparrow^{\frac{1}{2}N+s} \downarrow^{\frac{1}{2}N-s} \tag{1.13}$$

The coefficient of theterm in $\uparrow^{\frac{1}{2}N+s} \uparrow^{\frac{1}{2}N-s}$ is the number of states having $N\uparrow = 1/2\ N + s$ magnets up and $N\downarrow = 1/2\ N - s$ magnets down. This class of states has spin excess $N\uparrow - N\downarrow = 2s$ and net magnetic moment $2sm$. Let us denote the number of states in this class by $g(N, s)$, for a system of N magnets:

$$g(N, s) = \frac{N!}{\left(\frac{1}{2}N + s\right)!\left(\frac{1}{2}N - s\right)!} = \frac{N!}{N\uparrow !N\downarrow !} \tag{1.14}$$

Thus Eq. (1.13) is written as

$$(\uparrow + \downarrow)^N = \sum_{s=-\frac{1}{2}N}^{\frac{1}{2}N} g(N, s) \uparrow^{\frac{1}{2}N+s} \downarrow^{\frac{1}{2}N-s} \tag{1.15}$$

We shall call $g(N, s)$ the multiplicity function; it is the number of states having the same value of s. The reason for our definition emerges when a magnetic field is applied to the spin system: in a magnetic field, states of different values of s have different values of the energy, so that our g is equal to the multiplicity of an energy level in a magnetic field. Until we introduce a magnetic field, all states of the model system have the same energy, which may be taken as zero. Note from Eq. (1.15) that the total number of states is given by

$$\sum_{s=-\frac{1}{2}N}^{s=\frac{1}{2}N} g(N,s) = (1+1)^N = 2^N \tag{1.16}$$

Examples related to $g(N,s)$ for $N=10$ are given in Figure 1.6 and Figure 1.7. For a coin, "heads"could stand for "magnet down".

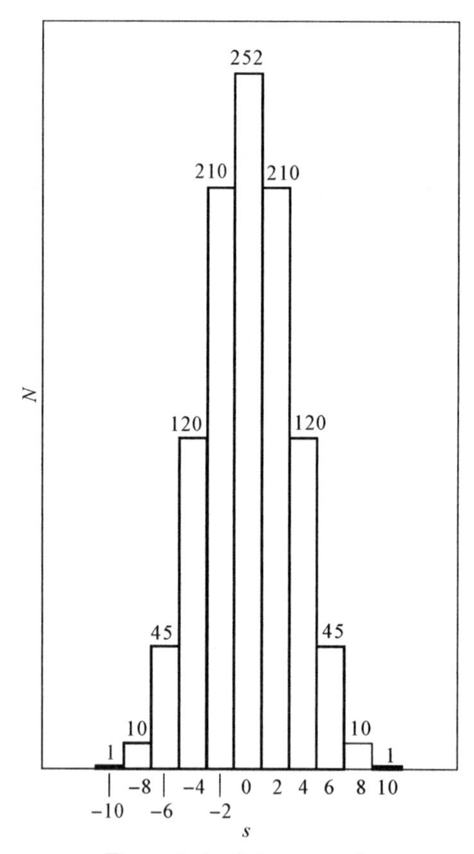

Figure 1.6 Spin excess $2s$

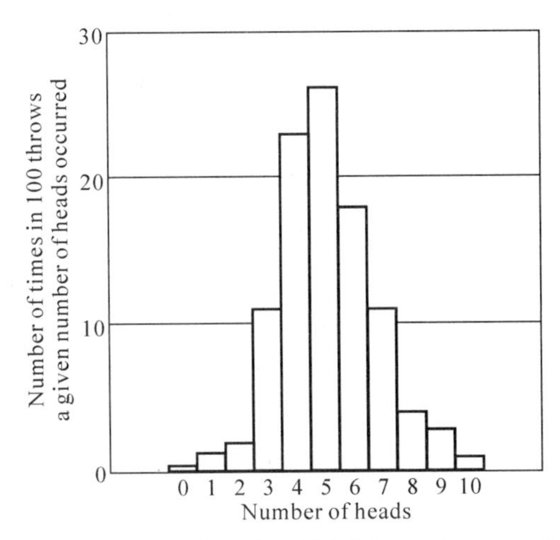

Figure 1.7 An experiment was done in which 10 pennies were thrown 100 times. The number of heads in each throw was recorded

Figure 1.6 Number of distinct arrangements of $5+s$ spins up and $5-s$ spins down. Figure 1.7 An experiment was done in which 10 pennies were thrown 100 times. Values of $g(N,s)$ are for $N=10$, where $2s$ is the spin excess $N\uparrow - N\downarrow$. The total number of states is

$$2^{10} = \sum_{s=-5}^{s} g(10,s)$$

The values of the $g's$ are taken from a table of the binomial coefficients.

1.3 Binary Alloy System

To illustrate that the exact nature of the two states on each site is irrelevant to the result, we consider an alternate system—an alloy crystal with N distinctsites, numbered from 1 through 12 in Figure 1.8.

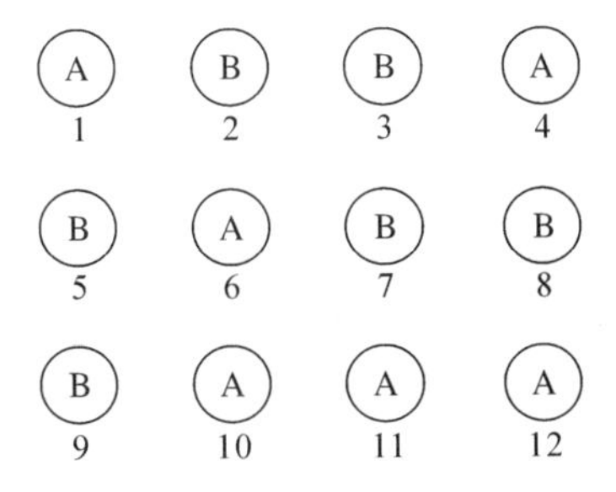

Figure 1.8 A binary alloy system of two chemical components A and B, whose atoms occupy distinct numbered sites

Each site is occupied by either an atom of chemical species A or an atom of chemical species B, with no provision for vacant sites. In brass, A could be copper and B zinc. In analogy to Eq. (1.2), a single state of the alloy system can be written as

$$A1B2B3A4B5A6B7B8B9A10A11A12\cdots \tag{1.17}$$

Every distinct stale of a binary alloy system on N sites is contained in the symbolic product of N factors:

$$(A_1+B_1)(A_2+B_2)(A_3+B_3)\cdots(A_N+B_N) \tag{1.18}$$

in analogy to Eq. (1.3). The average composition of a binary alloy isspeifed conventionally by the chemical formula $A_{1-x}B_x$, which mcans that out of a total of N atoms, the number of A atoms is $N_A=(1-x)N$ and the number of B atoms is $N_B=xN$. Here x lies between 0 and 1.

The symbolic expression

$$(A+B)^N=\sum_{t=0}^{N}\frac{N!}{(N-t)!t!}A^{N-t}B^t \tag{1.19}$$

is analogous to the result of Eq. (1.11). The coefficient of the term in $A^{N-t}B^t$ gives the number $g(N,t)$ of possible arrangements or states of $N-1$ atoms A and t atoms B on N sites:

$$g(N,t)=\frac{N!}{(N-t)!t!}=\frac{N!}{N_A!N_B!} \tag{1.20}$$

which is identical to the result of Eq. (1.14) for the spin model system, except for notation.

1.4 Sharpness of the Multiplicity Function

We know from common experience that systems held at constant temperature usually have well-defined properties; this stability of physical properties is a major prediction of thermal physics. The stability follows as a consequence of the exceedingly sharp peak in the multiplicity function and ofthe steep variation of that function away from the peak. We can show explicitly that for a very large system, the function $g(N,s)$ defined by Eq. (1.14) is peaked very sharply about the values $s=0$. We look for an approximation that allows us to examine the form of $g(N,s)$ versus s when $N\gg1$ and $|s|\ll N$. We cannot look up these values in tables: common tables of factorials do not go above $N=100$, and we may be interested in $N\approx10^{20}$, of the order of the number of atoms in a solid specimen big enough to be seen and felt. An approximation is clearly needed, and a good one is available.

It is convenient to work with $\ln g$. Except where otherwise specifed, all logarithms are understood to be log base e, writen here as ln. When you confront a very, very large number such as 2^N, where $N=10^{20}$, it is a simplification to look at the logarithm of the number. We take the logarithm of both sides of Eq. (1.14) to obtain

$$\ln g(N,s) = \ln N! - \ln\left(\frac{1}{2}N + s\right)! - \ln\left(\frac{1}{2}N - s\right)! \tag{1.21}$$

by virtue of the characteristic property of a product:

$$\ln xy = \ln x + \ln y,\ \ln\frac{x}{y} = \ln x - \ln y \tag{1.22}$$

With the notation

$$N\uparrow = \frac{1}{2}N + s,\ N\downarrow = \frac{1}{2}N - s \tag{1.23}$$

for thenumber of magnets up and down, Eq. (1.21) appears as

$$\ln g(N,s) = \ln N! - \ln N\uparrow ! - \ln N\downarrow ! \tag{1.24}$$

We evaluate the logarithm of $N!$ in Eq. (1.24) by use of the Stirling approximation, according to which

$$N! = \left(\frac{2\pi}{N}\right)^{\frac{1}{2}} N^N \exp\left[-N + \left(\frac{1}{12N}\right) + \cdots\right] \tag{1.25}$$

For $N \gg 1$. For sufficiently large N, the terms $\frac{1}{12N} + \cdots$ in the argument may be neglected in comparison with N. We take the logarithm of both sides of Eq. (1.25) to obtain.

$$\ln N\uparrow ! \approx \frac{1}{2}\ln 2\pi + \left(N + \frac{1}{2}\right)\ln N - N \tag{1.26}$$

Similarly

$$\ln N\uparrow ! \approx \frac{1}{2}\ln 2\pi + \left(N\uparrow + \frac{1}{2}\right)\ln N\uparrow - N\uparrow \tag{1.27}$$

$$\ln N\downarrow ! \approx \frac{1}{2}\ln 2\pi + \left(N\downarrow + \frac{1}{2}\right)9\ln N\downarrow - N\downarrow \tag{1.28}$$

After rearrangement of Eq. (1.26),

$$\ln N! \approx \frac{1}{2}\ln\left(\frac{2\pi}{N}\right) + \left(N\uparrow + \frac{1}{2} + N\downarrow + \frac{1}{2}\right)\ln N - (N\uparrow + N\downarrow) \tag{1.29}$$

where we have used $N = N\uparrow + N\downarrow$. We subtract Eq. (1.27) and Eq. (1.28) from Eq. (1.29) to obtain for Eq. (1.24):

$$\ln g \approx \frac{1}{2}\ln\left(\frac{1}{2\pi N}\right) - \left(N\uparrow + \frac{1}{2}\right)\ln\frac{N\uparrow}{N} - \left(N\downarrow + \frac{1}{2}\right)\ln\frac{N\downarrow}{N} \tag{1.30}$$

This may be simplified because

$$\ln\left(\frac{N\uparrow}{N}\right) = \ln\frac{1}{2}\left(1 + \frac{2s}{N}\right) = -\ln 2 + \ln\left(1 + \frac{2s}{N}\right) \approx$$

$$-\ln 2 + \left(\frac{2s}{N}\right) - \left(\frac{2s^2}{N^2}\right) \tag{1.31}$$

by virue of the expansion $\ln(1+x) = x - \frac{1}{2}x^2 + \cdots$, valid for $x \ll 1$. Similarly,

$$\ln\left(\frac{N\downarrow}{N}\right)=\ln\frac{1}{2}\left(1-\frac{2s}{N}\right)\approx-\ln 2-\left(\frac{2s}{N}\right)-\left(\frac{2s^2}{N^2}\right) \tag{1.32}$$

On substitution in Eq. (1.30) we obtain

$$\ln g\approx\frac{1}{2}\ln\left(\frac{2}{\pi N}\right)+N\ln 2-\frac{2s^2}{N} \tag{1.33}$$

We write this result as

$$g(N,s)\approx g(N,0)\exp\left(\frac{-2s^2}{N}\right) \tag{1.34}$$

where

$$g(N,0)\approx\left(\frac{2}{\pi N}\right)^{\frac{1}{2}}2^N \tag{1.35}$$

Such a distribution of values of s is called a Gaussian distribution. The integral* of Eq. (1.34) over the range $-\infty$ to $+\infty$ for s gives the correct value 2^N for the total number of states.

The exact value of $g(N, 0)$ is given by Eq. (1.14) with $s=0$:

$$g(N,0)=\frac{N!}{\left(\frac{1}{2}N\right)!\left(\frac{1}{2}N\right)!} \tag{1.36}$$

For $N=50$, the value of $g(50,0)$ is 1.264×10^{14}, from Eq. (1.36). The approximate value from Eq. (1.35) is 1.270×10^{14}. The distribution plotted in Figure 1.9 is centered in a maximum at $s=0$. When $s^2=1/2N$, the value of g is reduced to e^{-1} of the maximum value. That is, when

$$\frac{s}{N}=\left(\frac{1}{2N}\right)^{\frac{1}{2}} \tag{1.37}$$

the value of g is e^{-1} of $g(N, 0)$. The quantity $(1/2N)^{1/2}$ is thus a reasonable measure of the fractional width of the distribution. For $N\approx10^{22}$, the fractional width is of the order of 10^{-11}. When N is very large, the distribution is exceedingly sharply defined, in a relative sense. It is this sharp peak and the continued sharp variation of the multiplicity function far from the peak that will lead to a prcdiction that the physical propertics of systems in thermal equilibrium are well defined. We now consider one such property, the mean value of s^2.

---

* The replacement of a sum by an integral, such as $\sum_s(\cdots)$ by $\int(\cdots)ds$, usually does not introduce significant errors. For example, the ratio of

$$\sum_{s=0}^{N}s=\frac{1}{2}(N^2+N)\quad\text{to}\quad\int_0^N s\,ds=\frac{1}{2}N^2$$

is equal to $1+\left(\frac{1}{N}\right)$, which approaches 1 as N approaches ∞.

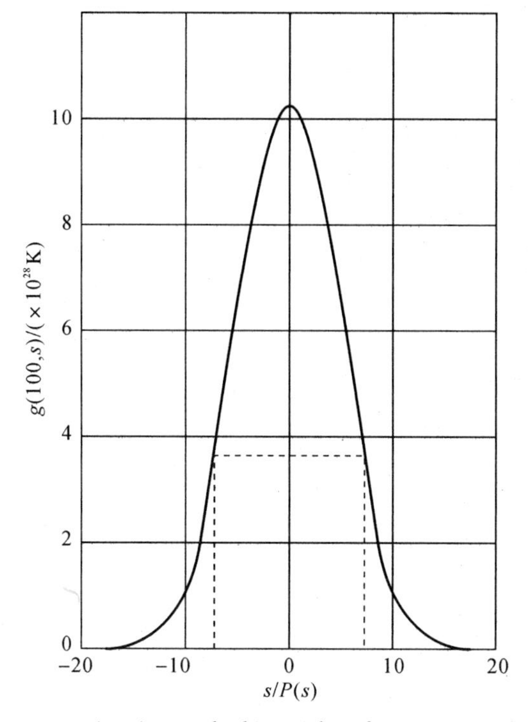

Figure 1.9 The Gaussian approximation to the binomial coeficients $g(100, s)$ plotted on a linear scale

1.5 Average Values

The average value, or mean value, of a function $f(s)$ taken over a probability distribution function $P(s)$ is defined as

$$\langle f\rangle = \sum_s f(s)P(s) \tag{1.38}$$

provided that the distribution function is normalized to unity:

$$\sum_s P(s) = 1 \tag{1.39}$$

The binomial distribution Eq. (1.14) has the property Eq. (1.16) that

$$\sum_s g(N, s) = 2^N \tag{1.40}$$

and is not normalized tounity. If all states are cqually probable, then $P(s)=g(N,s)/2^N$, and we have $\sum_s P(s) = 1$. The average of $f(s)$ over this distribution will be

$$\langle f\rangle = \sum_s f(s)P(N, s) \tag{1.41}$$

Consider the function $f(s)=s^2$. In the approximation that led to Eq. (1.34) and Eq. (1.35), we replace in Eq. (1.41) the sum $\sum$ over s by an integral $\int \cdots \mathrm{d}s$ between $-\infty$ and $+\infty$. Then

$$s^2 = \frac{(2/\pi N)^{\frac{1}{2}} \int_{-\infty}^{+\infty} s^2 \exp(-2s^2/N) \mathrm{d}s}{2^N} = (2/\pi N)^{1/2} (N/2)^{3/2} \int_{-\infty}^{+\infty} x^2 \mathrm{e}^{-x^2} \mathrm{d}x =$$

$$\left(\frac{2}{\pi N}\right)^{\frac{1}{2}} \left(\frac{N}{2}\right)^{\frac{3}{2}} (\pi/4)^{1/2}$$

whence

$$< s^2 > = \frac{1}{4} N, \quad < (2s)^2 > = N \tag{1.42}$$

The quntity $<(2s)^2>$ is the mean square spin excess. The root mean square spin excess is

$$< (2s)^2 >^{1/2} = \sqrt{N} \tag{1.43}$$

and the fractional fluctuation in $2s$ is defined as

$$\varepsilon \equiv \frac{< (2s)^2 >^{\frac{1}{2}}}{N} \frac{1}{\sqrt{N}} \tag{1.44}$$

The large N is, the smaller is the fractional fluctuation. This means that the central peak of the distribution function becomes relatively more sharply defined as the size of the system increases, the size being measured by the number of sites N. For 10^{20} particles, $\varepsilon = 10^{-10}$, which is very small.

1.6 Energy of the Binary Magnetic System

The thermal properties of the model system become physically relevant when the elementary magnets are placed in a magnetic field, for then the energies of the different states are no longer all equal. If the energy of the system is specified, then only the states having this energy may occur. The energy of interaction of a single magnetic moment m with a fixed external magnetic field B is

$$U = -mB \tag{1.45}$$

This is the potential energy of the magnet m in the field B.

For the model system of N elementary magnets, each with two allowed orientations in a uniform magnetic field B, the total potential energy U is

$$U = \sum_{i=1}^{N} u_i = -B \sum_{i=1}^{N} m_i = -2smB = -MB \tag{1.46}$$

using the expression M for the total magnetic moment $2sm$. In this example the spectrum of values of the energy U is discrete. We shall see later that a continuous or quasi-continuous spectrum will create no difficulty. Furthermore, the spacing between adjacent energy levels of this model is constant, as in Figure 1.10 . Constant spacing is a spcail fcature of the particular model, but this feature will not restrict the generality of the argument that is developed in the following sections.

The value of the energy for moments that interact only with the externalmagnctic field is completely determined by the value of s. This functional dependence is indicated by writing $U(s)$. Reversing a single moment lowers $2s$ by -2, lowers the total magnetic moment by $-2m$, and raises the energy by $2mB$. The energy difference between adjacent levels is denoted by $\Delta\varepsilon$, where

$$\Delta\varepsilon = U(s) - U(s+1) = 2mB \tag{1.47}$$

s	$U(s)/mB$	$g(s)$	$\ln g(s)$
−5 ————	+10	1	0
−4 ————	+8	10	2.30
−3 ————	+6	45	3.81
−2 ————	+4	120	4.79
−1 ————	+2	210	5.35
0 ————	0	252	5.53
+1 ————	−2	210	5.35
+2 ————	−4	120	4.79
+3 ————	−6	45	3.81
+4 ————	−8	10	2.30
+5 ————	−10	1	0

Figure 1.10 Energy levels of the model system of 10 magnetic moments m in a magnetic field B. The levels are labeled by their s values

Example: Multiplicity function for harmonic oscillator

The problem of the binary model system is the simplest problem for which an exact solution for the multiplicity function is known. Another exactly solvable problem is the harmonic oscillator, for which the solution was originally given by Max Planck. The original derivation is often felt to be not entirely simple. The beginning student need not worry about this derivation. The modern way to do the problem is given in Chapter 4 and is simple.

The quantum states of a harmonic oscillator have the energyeigenvalues

$$\varepsilon_s = s\hbar\omega \tag{1.48}$$

where the quantum number s is a positive integer or zero, and ω is the angular frequency of the oscillator. The number of states is infinite, and the multiplicity of each is one. Now consider a system of N such oscillators, all of the same frequency. We want to find the number of ways in which a given total excitation energy

$$\varepsilon = \sum_{i=1}^{N} s_i \hbar\omega = n\hbar\omega \tag{1.49}$$

can be distributed among the oscillators. That is, we want the multiplicity function $g(N, n)$ for the N oscillator . The oscillator multiplicity function is not the same as the spin multiplicity function found earlier.

We begin the analysis by going back to the multiplicity function for a single oscillator, for which $g(1,n)=1$ for all values of the quantum number s, here identical to n. To solve the problem of Eq. (1.52) below, we need a function to represent or generate the series

$$\sum_{n=0}^{\infty} g(1,n)\, t^n = \sum_{n=0}^{\infty} t^n \tag{1.50}$$

All $\sum$ run from 0 to ∞. Here t is just a temporary tool that will help us find the result of Eq. (1.52), but t does not appear in the final result . The answer is

$$\frac{1}{1-t} = \sum_{n=0}^{\infty} t^n \tag{1.51}$$

provided we assume $|t|<1$. For the problem of N oscillators, the generating function is

$$\left(\frac{1}{1-t}\right)^N = \left(\sum_{s=0}^{\infty} t^s\right)^N = \sum_{n=0}^{\infty} g(N,n) t^n \tag{1.52}$$

because the number of ways a term t^n can appear in the N-fold product is precisely the number of ordered ways in which the integer n can be formed as the sum of N non-negative integers.

We observe that

$$g(N,n) = \lim_{t\to 0} \frac{1}{n!}\left(\frac{\mathrm{d}}{\mathrm{d}t}\right)^n \sum_{s=0}^{\infty} g(N,s)\, t^s = \lim_{t\to 0} \frac{1}{n!}\left(\frac{\mathrm{d}}{\mathrm{d}t}\right)^n (1-t)^{-N} =$$

$$\frac{1}{n!} N(N+1)(N+2)\cdots(N+n-1) \tag{1.53}$$

Thus for the system of oscillators,

$$g(N,n) = \frac{(N+n-1)}{n!(N-1)!} \tag{1.54}$$

This result will be needed in solving a problem in the next chapter.

1.7 Summary

(1) The multiplicity function for a system of N magnets with spin excess $2s = N\uparrow - N\downarrow$ is

$$g(N,s) = \frac{N!}{\left(\frac{1}{2}N+s\right)!\left(\frac{1}{2}N-s\right)!} = \frac{N!}{N\uparrow!\,N\downarrow!}$$

In the limit $\frac{s}{N} \ll 1$, with $N \gg 1$, we have the Gaussian approximation:

$$g(N,s) \simeq \left(\frac{2}{\pi N}\right)^{\frac{1}{2}} 2^N \exp\left(\frac{-2s^2}{N}\right)$$

(2) If all states of the model spin system are equally likely, the average value of s^2 is

$$<s^2> = \frac{\int_{-\infty}^{\infty} s^2 g(N,s)\mathrm{d}s}{\int_{-\infty}^{\infty} g(N,s)\mathrm{d}s} = \frac{1}{4}N$$

in the Gaussian approximation.

(3) The fractional fluctuation of s^2 is defined as $\frac{<s^2>^{\frac{1}{2}}}{N}$ and is equal to $\frac{1}{2}N^{\frac{1}{2}}$.

(4) The energy of the model spin system in a state of spin excess $2s$ is

$$U(s) = -2smB$$

where m is the magnetic moment of one spin and B is the magnetic field.

Chapter 2 Temperature and Entropy

We start this chapter with a definition of probability that enables us to define the average value of a physical property of a system. We then consider systems in thermal equilibrium, the definition of entropy, and the definition of temperature. The second law of thermodynamics will appear as the law of increase of entropy. This chapter is perhaps the most abstract in the book. The chapters that follow will apply the concepts to physical problems.

2.1 Fundamental Assumption

The fundamental assumption of thermal physics is that a closed system is equally likely to be in any of the quantum states accessible to it. All accessible quantum states are assumed to be equally probable—there is no reason to prefer some accessible states over other accessible states.

A closed system will have constant energy, a constant number of particles, constant volume, and constant values of all external parameters that may influence the system, including gravitational, electric, and magnetic fields.

A quantum state is accessible if its properties are compatible with the physical specification of the system: the energy of the state must be in the range within which the energy of the system is specified, and the number of particles must be in the range within which the number of particles is specified. With large systems we can never know either of these exactly, but it will suffice to have $\Delta U/U \ll 1$ and $\delta N/N \ll 1$.

Unusual properties of a system may sometimes make it impossible for certain states to be accessible during the time the system is under observation. For example, the states of the crystalline form of SiO_2, are inaccessible at low temperatures in any observation that starts with the glassy or amorphous form: fused silica will not convert to quartz in our lifetime in a low-temperature experiment. You will recognize many exclusions of this type by common sense. We treat all quantum states as accessible unless they are excluded by the specification of the system (see Figure 2.1) and the time scale of the measurement process. States that are not accessible are said to have zero probability.

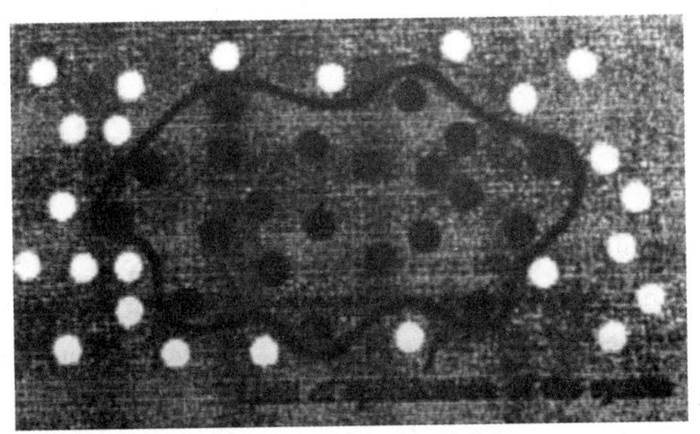

Figure 2.1 A purely symbolic diagram: each solid spot represents an accessible quantum state of a closed system

Of course, it is possible to specify the configuration of a closed system to a point that its statistical properties as such are of no interest. If we specify that the system is exactly in a stationary quantum states, no statistical aspect is left in the problem.

2.2 Probability

Suppose we have a closed system that we know is equally likely to be in any of the g accessible quantum states. Let s be a general state label (and not one-half the spin excess). The probability $P(s)$ of finding the system in this state is

$$P(s) = 1/g \tag{2.1}$$

if the state s is accessible and $P(s)=0$ otherwise, consistent with the fundamental assumption. We shall be concerned later with systems that are not closed, for which the energy U and particle number N may vary. For these systems $P(s)$ will not be a constant as in Eq. (2.1), but will have a functional dependence on U and on N.

The sum $P(s)$ of the probability over all states is always equal to unity,ecause the total probability that the system is in some state is unity:

$$\sum P(s) = 1 \tag{2.2}$$

The probabilities defined by Eq. (2.1) lead to the definition of the average value of any physical property. Suppose that the physical property X has the value $X(s)$ when the system is in the state s. Here X might denote magnetic moment, energy, square of the energy, charge density near a point r, or any property that can be observed when the system is in a quantum state. Then the average of the observe actions of the quantity X taken over a system described by the probabilities $P(s)$ is

$$\langle X \rangle = \sum X(s)P(s) \tag{2.3}$$

This equation defines the average value of X. Here $P(s)$ is the probability that the system is in the state s. The angular brackets<…> are used to denote average value.

For a closed system, the average value of X is

$$\langle X\rangle = \sum X(s)(1/g) \tag{2.4}$$

because now all y accessible states are equally likely, with $P(s)=1/g$. The average in Eq. (2.4) is an elementary example of what may be called an ensemble average: we imagine g similar systems, one in each accessible quantum state. Such a group of systems constructed alike is called an ensemble of systems. The average of any property over the group is called the ensemble average of that property.

An ensemble of systems is composed of many systems, all constructed alike. Each system in the ensemble is a replica of the actual system in one of the quantum states accessible to the system. If there are g accessible states, then there will be y systems in the ensemble, one system for each state. Each system in the ensemble is equivalent for all practical purposes to the actual system. Each system satisfies all external requirements placed on the original system and in this sense is "just as good" as the actual system. Every quantum state accessible to the actual system is represented in the ensemble by one system by one system in a stationary quantum state, as in Figure 2.2.

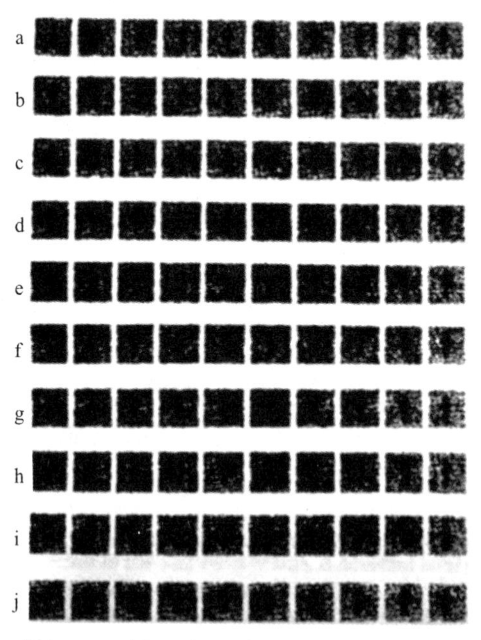

Figure 2.2 This ensemble a through j represents a system of 10 spins with energy $-8mB$ and spin excess $2s=8$

We assume that the ensemble represents the real system—this is implied in the fundamental assumption.

Example: Construction of an ensemble

We construct in Figure 2. 3 an ensemble to represent a closed system of five spins, each system with spin excess $2s=1$. The energy of each in a magnetic field is $-mB$(Do not confuse the use of s in spin excess with our frequent use of s as a state index or label.). Each system represents one of the multiples of quantum states at this energy.. The number of such states is given by the multiplicity function of Eq. (1. 14):

$$g\left(5,\frac{1}{2}\right)=\frac{5!}{3!2!}=10$$

The 10 systems shown in Figure 2. 3 make up the ensemble.

If the energy in the magnetic field were such that $2s=5$, then a single system comprises the ensemble, as in Figure 2. 4. In zero magnetic field, all energies of all $2^N=2^5=32$ states are equal. and the new ensemble must represent 32 systems, of which 1 system has $2s=5$; 5 systems have $2s=3$; 10 systems have $2s=1$;10 systems have $2s=-1$; 5 systems have $2s=-3$; and 1 system has $2s=-5$.

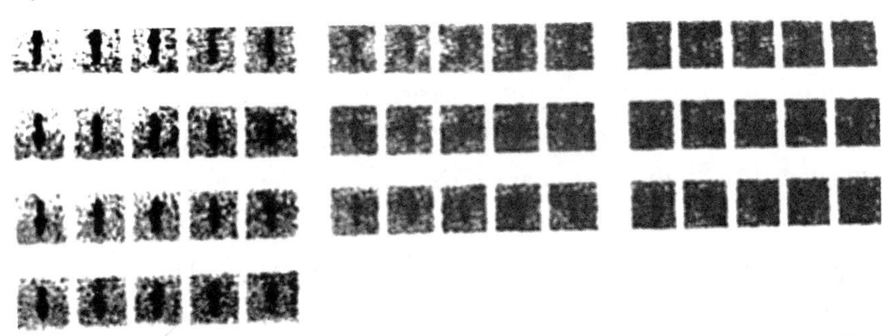

Figure 2. 3 The ensemble represents a system with $N=5$ spins and spin excess $2s=1$

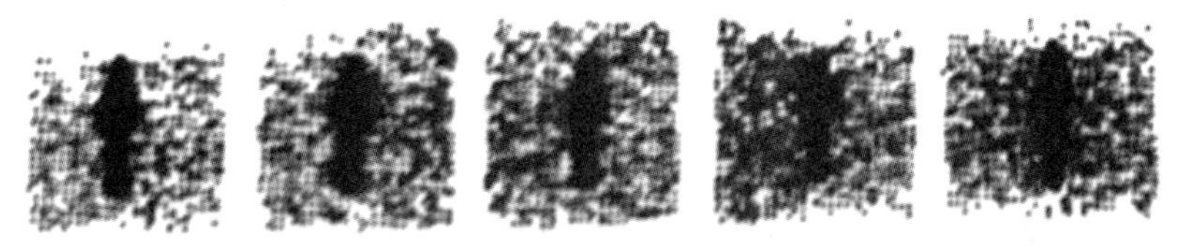

Figure 2. 4 With $N=5$ and $2s=5$, a single system may represent the ensemble. This is not a typical situation

2. 3 Most Probable Configuration

Let two systems S_1 and S_2, be brought into contact so that energy can be transfer from one to the other. This is called thermal contact (see Figure 2.5). The two systems in contact form a larger closed system $S=S_1+S_2$ with constant energy $U=U_1+U_2$. What determines whether there will be a net flow of energy from one system to another? The answer leads to the concept of temperature. The direction of energy flow is not simply a matter of whether the energy of one system is greater than the energy of the other, because the systems can be different in size and constitution. A constant total energy can be shared in many ways between two systems.

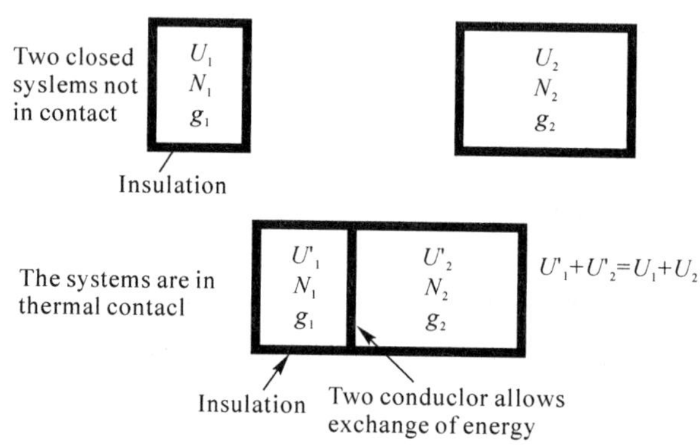

Figure 2.5　Establishment of themal contact between two systems(S_1 and S_2)

The most probable division of the total energy is that for which the combined system has S_2 the maximum number of accessible states. We shall enumerate the accessible states of two model systems and then study what characterizes the systems when in thermal contact. We first solve in detail the problem of thermal contact between two spin systems, S_1 and S_2, in a magnetic field which is introduced in order to define the energy. The numbers of spins N_1, N_2 may be different, and the values of the spin excess 2_{s_1}, 2_{s_2} may be different for the two systems. All spins have magnetic moment m. The actual exchange of energy might take place via some weak (residual) coupling between the spins near the interface between the two systems. We keep N_1, N_2 constant, but the values of the spin excess are allowed to change. The spin excess of a state of the combined system will be denoted by $2s$, where $s=s_1+s_2$. The energy of the combined system is directly proportional to the total spin excess:

$$U(s)=U_1(s_1)+U_2(s_2)=-2mB(s_1+s_2)=-2mBs \tag{2.5}$$

The total number of particles is $N=N_1+N_2$.

We assume that the energy splittings between adjacent energy levels are equal $102mB$ in both systems, so that the magnetic energy given up by system S_1 when one spin is reversed can be taken up by the reversal of one spin of system S_2 in the opposite sense. Any large physical system will have enough diverse modes of energy so that energy exchange with another system is always possible. The value of $s=s_1+s_2$ is constant because the total energy is constant, but when the two systems are brought into thermal contact a redistribution is permitted in the separate values of s_1, s_2 and thus in the energies U_1, U_2.

The multiplicity function $g(N,s)$ of the combined system 8 is related to the product of the multiplicity functions of the individual systems 8, by the relation:

$$g(N,s) = \sum_{s_1} g_1(N_1,s_1)g_2(N_2,s-s_1) \tag{2.6}$$

where the multiplicity functions g_1, g_2 are given by expressions of the form of Eq. (1.14). The range of s_1 in the summation is from $-1/2N_1$ to $1/2N_1$, if $N_1 < N_2$. To see how Eq. (2.6) comes about, consider first that configuration of the combined system for which the first system has spin excess $2s_2$ and the second system has spin excess $2s_2$. A configuration is defined as the set of all states with specified values of s_1 and s_2. The first system has $g_1(N_1, s_1)$ accessible states, each of which may occur together with any of the $g_2(N_2,s_2)$ accessible states of the second system. The total number of states in one configuration of the combined system is given by the product $g_1(N_1, s_1)$ $g_2(N_2,s_2)$ of the multiplicity functions of S_1 and S_2. Because $s_2 = s - s_1$, the product of the $g's$ may be written as

$$g_1(N_1,s_1)g_2(N_2,s-s_1) \tag{2.7}$$

This product forms one term of the sum of Eq. (2.6).

Different configurations of the combined system are characterized by different values of s_1 We sum over all possible values of s_1 to obtain the total number of states of all the configurations with fixed s or fixed energy. We thus obtain Eq. (2.6), where $g(N,s)$ is the number of accessible states of the combined system. In the sum we hold s, N_1, and N_2 constant, as part of the specification of thermal contact.

The result of Eq. (2.6) is a sum of products of the form Eq. (2.7). Such a product will be a maximum for some value of s_1, say $\hat{s}_1$, to be read as "s_1 hat" or "s_1 caret". The configuration for which $g_1 g_2$ is a maximum is called the most probable configuration; the number of states in it is

$$g_1(N_1,\hat{s}_1)g_2(N_2,s-\hat{s}_1) \tag{2.8}$$

If the system are large, the maximum with respect to changes in s_1 will be extremely sharp, as in Figure 2.6. A relatively small number of configurations will dominate the statistical properties of the combined system, The most probable configuration alone will describe many of these properties.

Such a sharp maximum is a property of every realistic type of large system for which exact solutions are available; we postulate that it is a general property of all large systems. From the sharpness property it follows that fluctuations about the most probable configuration are small, in a sense that we will define.

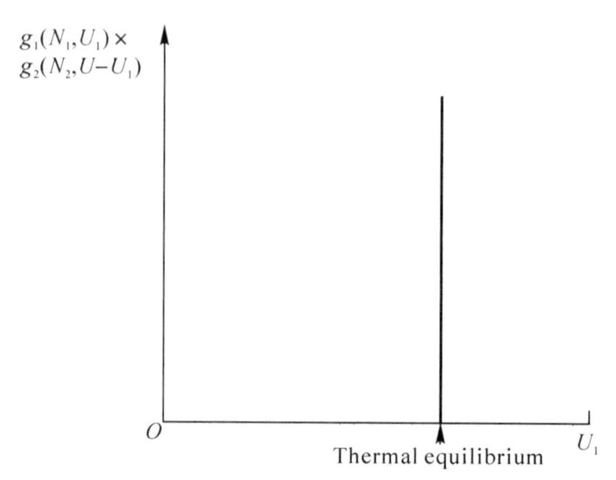

Figure 2. 6 Schematic representation of the dependence of the configuration multiplicity on the division of the total energy between two system

The important result follows that the values of the average physical properties of a large system in thermal contact with another large system are accurately described by the properties of the most probable configuration, the configuration for which the number of accessible states is a maximum. Such average values (used in either of these two senses) are called thermal equilibrium values.

Because of the sharp maximum, we may replace the average of a physical quantity over all accessible configurations Eq. (2. 6) by an average over only the most probable configuration Eq. (2. 8). In the example below we estimate the error involved in such a replacement and find the error to be negligible.

Example: Two spin systems in thermal contact

We investigate for the model spin system the sharpness of the product Eq. (2. 7) near the maximum Eq. (2. 8) as follows. We form I the product of the multiplicity functions for $g_1(N_1, s_1)$ and $g_2(N_2, s_2)$. both of the form of Eq. (1. 34):

$$g_1(N_1,s_1)g_2(N_2,s_2) = g_1(0)g_2(0)\exp\left(-\frac{2s_1{}^2}{N_1}-\frac{2s_2{}^2}{N_2}\right) \tag{2.9}$$

where $g_1(0)$ denotes $g_1(N_1, 0)$ and $g_2(0)$ denotes $g_2(N_1, 0)$, we replace s_2 by $s-s_1$:

$$g_1(N_1,s_1)g_2(N_2,s-s_1) = g_1(0)g_2(0)\exp\left[-\frac{2s_1{}^2}{N_1}-\frac{2(s-s_1)^2}{N_2}\right] \tag{2.10}$$

This product gives the number of states accessible to the combined system when the spin excess of the combined system is $2s$, and the spin excess of the first system is $2s_1$.

We find the maximum value of Eq. (2. 10) as a function of s_1 when the total spin excess $2s$ is held constant; that is, when the energy of the combined systems is constant. It is convenient to use the property that the maximum of $\ln y(x)$ occurs at the same value of x as the maximum of $y(x)$. The calculation can be done either way. From Eq. (2.10),

$$\ln g_1(N_1, s_1) g_2(N_2, s-s_1) = \ln g_1(0) g_2(0) - \frac{2s_1{}^2}{N_1} - \frac{2(s-s_1)^2}{N_2} \tag{2.11}$$

This quantity is an extremum when the first derivative with respect to s_1 is zero. An extremum may be a maximum, a minimum, or a point or inflection. The extremum is a maximum if the second derivative of the function is negative, so that the curve bend downward.

At the extremum the first derivative is

$$\frac{\partial}{\partial s_1}[\ln g_1(N_1, s_1) g_2(N_2, s-s_1)] = -\frac{4s_1}{N_1} + \frac{4(s-s_1)}{N_2} \tag{2.12}$$

where N_1, N_2, and s are held constant as s_1 is varied. The second derivative $\frac{\partial^2}{\partial s_1{}^2}$ of Eq. (2.11) is

$$-4\left(\frac{1}{N_1} + \frac{1}{N_2}\right)$$

The product function of two Gaussian functions is always a Gaussian, and is negative, so that the extremum is a maximum. Thus the most probable configuration of the combined system is that for which Eq. (2.12) is satisfied:

$$\frac{s_1}{N_1} = \frac{s-s_1}{N_2} = \frac{s_2}{N_2} \tag{2.13}$$

The two systems are inequilibrium with respect to interchange of energy when the fractional spin excess of system 1 is equal to the fractional spin excess of system 2.

We prove that nearly all the accessible states of the combined systems satisfy or very nearly satisfy Eq. (2.13). If $\hat{s}$ and $\hat{s}_2$ denote the values of s_1 and s_2 at the maximum, then Eq. (2.13) is written as

$$\frac{\hat{s}_1}{N_1} = \frac{s}{N_2} = \frac{\hat{s}_2}{N} \tag{2.14}$$

To find the number of states in the most probable configuration, we insert Eq. (2.14) in Eq. (2.9) to obtain

$$(g_1 g_2)_{\max} = g_1(\hat{s}_1) g_2(s-\hat{s}_1) = g_1(0) g_2(0) \exp(-2s^2/N) \tag{2.15}$$

To investigate the sharpness of the maximum of $g_1 g_2$ at a given value of s, introduce δ such that

$$s_1 = \hat{s}_1 + \delta, \quad s_2 = \hat{s}_2 + \delta \tag{2.16}$$

here, δ measures the deviation of s_1, s_2 at the maximum of $g_1 g_2$. Square s_1, s_2 to from

$$s_1{}^2 = \hat{s}_1{}^2 + 2\hat{s}_1\delta + \delta^2, \quad s_2{}^2 = \hat{s}_2{}^2 - 2\hat{s}_2\delta + \delta^2$$

Which we substitute in Eq. (2.9) and Eq. (2.15) to obtain the number of states:

$$g_1(N_1, s_1) g_2(N_2, s-s_1) = (g_1 g_2)_{\max} \exp\left(-\frac{4\hat{s}_1\delta}{N_1} - \frac{2\delta^2}{N_1} + \frac{4\hat{s}_2\delta}{N_2} - \frac{2\delta^2}{N_2}\right)$$

We know from Eq. (2.14) that $\frac{\hat{s}_1}{N_1}=\frac{\hat{s}_2}{N_2}$, so that the number of states in a configuration of deviation δ from equilibrium is

$$g_1(N_1,\hat{s}_1+\delta)g_2(N_2,\hat{s}_2-\delta) = (g_1 g_2)_{\max}\exp\left(-\frac{2\delta^2}{N_1}-\frac{2\delta^2}{N_2}\right) \tag{2.17}$$

As a numerical example in which the fractional deviation from equilibrium is very small, let $N_1=N_2=10^{22}$ and $\delta=10^{12}$ that is $\delta/N_1=10^{-10}$. Then $2\delta^2$, $N_1=200$, and the product $g_1 g_2$ is reduced to $e^{-400}\approx 10^{-174}$ of its maximum value. This is an extremely large reduction, so that $g_1 g_2$ is truly a very sharply peaked function of s_1. The probability that the tractional deviation will be 10^{-10} or larger is found by integrating Eq. (2.17) from $\delta=10^{12}$ out to a value of the order of s or of N. Thereby including the area under the wings of the probability distribution. This is the subject of Problem 6. An upper limit to the integrated probability is given by $N\times 10^{-174}=10^{-152}$, still very small. When two systems are in thermal contact. the values of s_1, s_2 that occur most often will be very close to the values of $\hat{s}_1\hat{s}_2$ for which the product $g_1 g_2$ is a maximum. It is extremely rare to find systems with values of s_1, s_2 perceptibly different from $\hat{s}_1\hat{s}_2$.

What does it mean to say that the probability of finding the system with a fractional deviation larger than $\delta/N_1=10^{-10}$ is only 10^{-152} of the probability of finding the system in equilibrium? We mean that the system will never be found with a deviation as much as I part in 10^{10}, small as this deviation seems. We would have to sample 10^{152} similar systems to have a reasonable chance of success in such an experiment. If we sample one system every 10^{-12} s, which is pretty fast work, we would have to sample for 10^{140} s. The age of the universe is only 10^{18} s. Therefore we say with great surety that the deviation described will never be observed. The estimate is rough, but the message is correct. The quotation from Boltzmann given at the beginning of this chapter is relevant here.

We may expect to observe substantial fractional deviations only in the properties of a small system in thermal contact with a large system or reservoir. The energy of a small system, say a system of 10 spins, in thermal contact with a large reservoir may undergo fluctuations that are large in a fractional sense, as have been observed in experiments on the Brownian motion of small particles in suspension in liquids. The average energy of a small system in contact with a large system can always be determined accurately by observations at one time on a large number of identical small systems or by observations on one small system over a long period of time.

2.4 Thermal Equilibrium

The result for the number of accessible states of two model spin systems in thermal contact may be generalized to any two systems in thermal contact, with constant total energy $U=U_1+U_2$. By direct extension of the earlier argument, the multiplicity $g(N,U)$ of the combined system is

$$g(N,U) = \sum_{U_1} g_1(N_1,U_1)g_2(N_2,U-U_1) \tag{2.18}$$

summed over all values of $U_1 \leqslant U$. Here $g_1(N_1,U_1)$ is the number of accessible states of system 1 at energy U_1. A configuration of the combined system is specified by the value of U_1, together with the constants U, N_1, N_2. The number of acessible states in a configuration is the product $(N_1,U_1)g_2(N_2,U-U_1)$. The sum over all configurations gives $g(N,U)$.

The largest term in the sum in Eq. (2.18) governs the properties of the total system in thermal equilibrium. For an extremum it is necessary that the differential of $g(N,U)$ be zero for an infinitesimal exchange of energy:

$$\mathrm{d}g = \left(\frac{\partial g_1}{\partial U_1}\right)_{N_1} g_2 \mathrm{d}U_1 + g_1 \left(\frac{\partial g_2}{\partial U_2}\right)_{N_2} \mathrm{d}U_2 = 0, \ \mathrm{d}U_1 + \mathrm{d}U_2 = 0 \tag{2.19}$$

We divide by $g_1 g_2$ and use the result $\mathrm{d}U_1 = -\mathrm{d}U_2$ to obtain the thermal equilibrium condition:

$$\frac{1}{g_1}\left(\frac{\partial g_1}{\partial U_1}\right)_{N_1} = \frac{1}{g_2}\left(\frac{\partial g_2}{\partial U_2}\right)_{N_2} \tag{2.20a}$$

which we may write as

$$\left(\frac{\partial \ln g_1}{\partial U_1}\right)_{N_1} = \left(\frac{\partial \ln g_2}{\partial U_2}\right)_{N_2} \tag{2.20b}$$

We define the quantity σ, called the entropy, by

$$\sigma(N,U) = \ln g(N,U) \tag{2.21}$$

where σ is the Greek letter sigma. We now write Eq. (2.20) in the final form:

$$\left(\frac{\partial \sigma_1}{\partial U_1}\right)_{N_1} = \left(\frac{\partial \sigma_2}{\partial U_2}\right)_{N_2} \tag{2.22}$$

The notation

$$\left(\frac{\partial g_1}{\partial U_1}\right)_{N_1}$$

means that N_1 is held constant in the differentiation of $g_1(N_1,U_1)$ with respect to U_1. That is, the partial derivative with respect to U_1 is defined as

$$\left(\frac{\partial g_1}{\partial U_1}\right)_{N_1} = \lim_{\Delta U_1 \to 0} \frac{g_1(N_1,U_1+\Delta U_1) - g_1(N_1,U_1)}{\Delta U_1}$$

For example, if $g(x,y)=3x^4 y$, then $\partial g/\partial x = 12x^3 y$ and $\partial g/\partial y = 3x^4$.

This is the condition for thermal equilibrium for two systems in thermal contact. Here N_1 and N_2 may symbolize not only the numbers of particles, but all constraints on the systems.

2.5 Temperature

The last Eq. (2.22) leads us immediately to the concept of temperature. We know the everyday rule: in thermal equilibrium the temperatures of the two systems are equal:

$$T_1 = T_2 \tag{2.23}$$

This rule must be equivalent to Eq. (2.22), so that T must be a function of $(\partial\sigma/\partial U)_N$. If T denotes the absolute temperature in Kelvin, this function is simply the inverse relationship:

$$\frac{1}{T} = k_B\left(\frac{\partial\sigma}{\partial U}\right)_N \tag{2.24}$$

The proportionality constant k_B is a universal constant called the Boltzman constant. As determined experimentally,

$$k_B = 1.381 \times 10^{-23}\ \mathrm{J/K} = 1.381 \times 10^{-16}\ \mathrm{erg/K} \tag{2.25}$$

We prefer to use a more natural temperature scale: we define the fundamental temperature t by

$$\frac{1}{\tau} = \left(\frac{\partial\sigma}{\partial U}\right)_N \tag{2.26}$$

The temperature differs from the Kelvin temperature by the scale factor, k_B:

$$\tau = k_B T \tag{2.27}$$

Because σ is a pure number, the fundamental temperature τ has the dimensions of energy. We can use as a temperature scale the energy scale, in whatever unit may be employed for the latter-joule or erg. This procedure is much simpler than the introduction of the Kelvin scale in which the unit of temperature is arbitrarily selected so that the triple point of water is exactly 273.16 K. The triple point of water is the unique temperature at which water, ice, and water vapor coexist.

Historically, the conventional scale dates from an age in which it was possible to build accurate thermometers even though the relation of temperature to quantum states was as yet not understood. Even at present, it is still possible to measure temperatures with thermometers calibrated in kelvin to a higher precision than the accuracy with which the conversion factor k_B itself is known about 32 parts per million.

Comment. In Eq. (2.26) we defined the reciprocal of τ as the partial derivative $(\partial\sigma/\partial U)_N$. It is permissible to take the reciprocal of both sides to write

$$\tau = (\partial U/\partial\sigma)_N \tag{2.28}$$

The two expressions of Eq. (2.26) and Eq. (2.28) have a slightly different meaning. In

Eq. (2.26), the entropy σ was given as a function of the independent variables U and N as $\sigma=\sigma(U,N)$. Hence τ determined from Eq. (2.26) has the same independent variable, $\tau=\tau(U, N)$. In Eq. (2.28), however, differentiation of U with respect to σ with N constant implies $U=U(\sigma,N)$, so that $\tau=\tau(\sigma, N)$. The definition of temperature is the same in both cases but it is expressed as a function of different independent variables. The question : What are the independent variable? arise frequently in thermal physics because in some experiments we control some variables, and in other experiments we control other variables.

2.6 Entropy

The quantity $\sigma=\ln g$ was introduced in Eq. (2.21) as the entropy of the system: the entropy is defined as the logarithm of the number of states to the system. As defined, the entropy S is defined by

$$\frac{1}{T}=\left(\frac{\partial S}{\partial U}\right)_N \tag{2.29}$$

As a consequence of Eq. (2.24), we see that S and σ are connected by a scale factor:

$$S=k_B\sigma \tag{2.30}$$

We will call S the conventional entropy.

The more states that are accessible, the greater the entropy. In the definition of $\sigma(N, U)$ we have indicated a functional dependence of the entropy on the number of particles in the system and on the energy of the system. The entropy may depend on additional independent variables: the entropy of a gas (see Chapter 3) depends on the volume.

In the early history of thermal physics the physical significance of the entropy was not known. Thus the author of the article on thermodynamics in the Encyclopaedia Biannica, 11th ed. (1905), wrote: "The utility of the conception of entropy is limited by the fact that it does not correspond directly to any directly measurable physical property, but is merely a mathematical function of the definition of absolute temperature." We now know what absolute physical property the entropy measures. An example of the comparison of the experimental determination and theoretical calculation of the entropy is discussed in Chapter 6.

Consider the total entropy change $\Delta\sigma$ when we remove a positive amount of energy ΔU from 1 and add the same amount of energy to 2, as in Figure 2.7.

The total entropy change is

$$\Delta\sigma=\left(\frac{\partial\sigma_1}{\partial U_1}\right)_{N_1}(-\Delta U)+\left(\frac{\partial\sigma_2}{\partial U_2}\right)_{N_2}(\Delta U)=\left(-\frac{1}{\tau_1}+\frac{1}{\tau_2}\right)\Delta U \tag{2.31}$$

When $\tau_1>\tau_2$ the quantity in parentheses on the right-hand side is positive, So that the total change of entropy is positive when the direction of energy flow is from the system with the higher temperature to the system with the lower temperature.

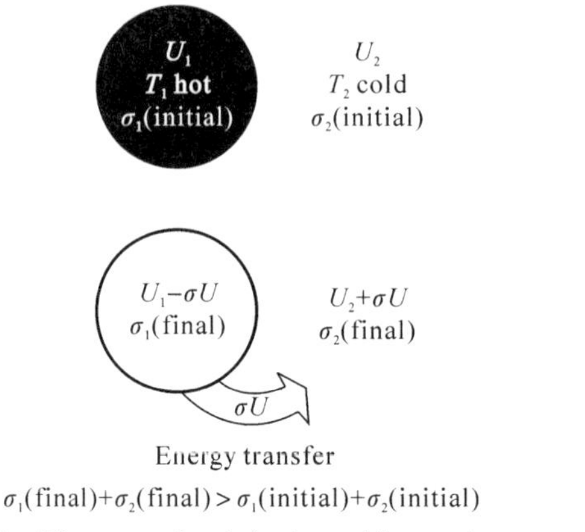

Figure 2.7 The example of the law of increasing entropy

Example: Entropy increase on heat flow

This example makes use of the reader's previous familiarity with heat and specific heat.

(1) Let a 10 g specimen of copper at a temperature of 350 K be placed in thermal contact with an identical specimen at a temperature of 290 K. Let us find the quantity of energy transferred when the two specimens are placed in contact and come to equilibrium at the final temperature T. The specific heat of metallic copper over the temperature range 15℃ to 100℃ is approximately 0.389 $\mathrm{J \cdot g^{-1} \cdot K^{-1}}$, according to a standard handbook.

The energy increase of the second specimen is equal to the energy loss of the first; thus the energy increase of the second specimen is, in joules,

$$\Delta U = (3.89\ \mathrm{J \cdot K^{-1}})(T_f - 290\ \mathrm{K}) = (3.89\ \mathrm{J \cdot K^{-1}})(350\ \mathrm{K} - T_f)$$

where the temperatures are in Kelvin. The final temperature after contact is

$$T_f = \frac{1}{2}(350 + 290)\mathrm{K} = 320\ \mathrm{K}$$

Thus

$$\Delta U_1 = (3.89\ \mathrm{J \cdot K^{-1}})(-30\ \mathrm{K}) = -11.7\ \mathrm{J}$$

And

$$\Delta U_2 = -\Delta U_1 = 11.7\ \mathrm{J}$$

(2) What is the change of entropy of the two specimens when a transfer of 0.1 J has taken place, almost immediately after initial contact? Notice that this transfer is a small fraction of the final energy transfer as calculated above. Because the energy transfer considered is small, we may suppose the specimens are approximately at their initial temperatures of 350 K and 290 K. The entropy of the first body is decreased by

$$\Delta S_1 = \frac{-0.1\text{ J}}{350\text{ K}} = -2.86 \times 10^{-4}\text{J} \cdot \text{K}^{-1}$$

The entropy of the second body is increased by

$$\left.\begin{aligned} &\Delta S_2 = \frac{0.1\text{ J}}{290\text{ K}} = 3.45 \times 10^{-4}\text{J} \cdot \text{K}^{-1} \\ &\Delta S_1 + \Delta S_2 = (-2.86 + 3.45) \times 10^{-4}\text{J} \cdot \text{K}^{-1} = 0.59 \times 10^{-4}\text{J} \cdot \text{K}^{-1} \\ &\Delta\sigma = \frac{0.59 \times 10^{-4}}{k_B} = \frac{0.59 \times 10^{-4}\text{J} \cdot \text{K}^{-1}}{1.38 \times 10^{-23}\text{ J} \cdot \text{K}^{-1}} = 0.43 \times 10^{19} \end{aligned}\right\} \quad (2.32)$$

where k_B is the Bolumann constant. This result means that the number of accssible states of the two systems increases by the factor $\exp(\Delta\sigma) = \exp(0.43 \times 10^{19})$.

2.7 Law of Increase of Entropy

We can show that the total entropy always increases when two systems are brought into thermal contact. We have just demonstrated this in a special case. If the total energy $U = U_1 + U_2$ is constant, the total multiplicity after the systems are in thermal contact is

$$g(U) = \sum_{U_1} g_1(U_1) g_2(U - U_1) \quad (2.33)$$

by Eq. (2.18). This expression contains the term $g_1(U_{10})\, g_2(U - U_{10})$ for the initial multiplicity before contact and many other terms besides. Here U_{10} is the initial energy of system 1 and $U - U_{10}$ is the initial energy of system 2. Because all terms in Eq. (2.33) are positive numbers, the multiplicity is always increased by establishment of thermal contact between two systems. This is a proof of the law of increase of entropy for a well defined operation.

The significant effect of contact, the effect that stands out even after taking the logarithm of the multiplicity is not just that the number of terms in the summation is large, but that the largest single term in the summation may be very much larger than the initial multiplicity. That is

$$(g_1 g_2)_{\max} = g_1(\hat{U}_1) g_2(U - \hat{U}_1) \quad (2.34)$$

may be very much larger than the initial term: $g_1(U_{10}) g_2(U - U_{10})$.

Here $\hat{U}$, denotes the value of U_1 for which the product $g_1 g_2$ is a maximum. The essential effect is that the systems after contact evolve from their initial configurations to their final configurations. The fundamental assumption implies that evolution in this operation will always take place, with all accessible final states equally probable.

The statement

$$\sigma_{\text{final}} \approx \ln(g_1 g_2)_{\max} \geqslant \sigma_{\text{initial}} = \ln(g_1 g_2)_0 \quad (2.35)$$

is a statement of the law of increase of entropy: the entropy of a closed system tends to remain constant or to increase when a constraint internal to the system is removed. The operation of establishing thermal contact is equivalent to there moval of the constraint that U_1, U_2 each be constant; after contact only U_1+U_2 need be constant.

The evolution of the combined system towards the final thermal equilibrium configuration takes a certain time. If we separate the two systems before they reach this configuration, we will obtain an intermediate configuration with intermediate energies and an intermediate entropy. It is therefore meaningful to view the entropy as a function of the time that has elapsed since removal of the constraint, called the time of evolution in Figure 2. 8.

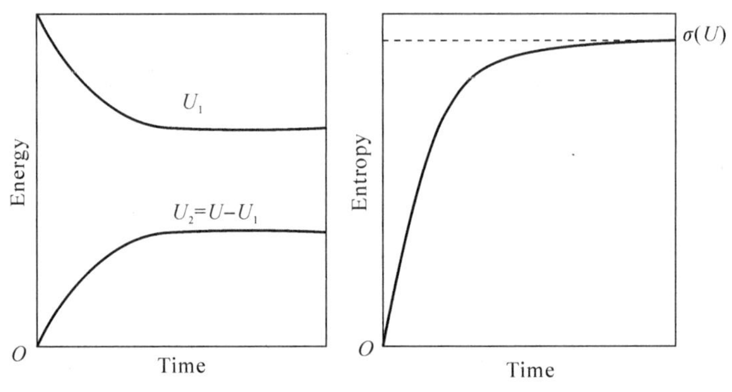

Figure 2. 8 A system with two parts

Processes that tend to increase the entropy of a system are shown in Figure 2. 9; the arguments in support of each process will be developed in the chapters that follow.

For a large system in thermal contact with another large system there will never occur spontaneously significant differences between the actual value of the entropy and the value of the entropy of the most probable configuration of the system. We showed this for the model spin system in the argument following Eq. (2. 17); we used "never" in the sense of not once in the entire age of the universe, 10^{18} s We can only find a significant difference between the actual entropy and thentropy of the most probable configuration of the macroscopic system very shortly after we have changed the nature of the contact between two systems, which implies that we had prepared the system initially in some special way. Special preparation could consist of lining up all the spins in one system parallel to one another or collecting all the molecules in the air of the room into the system formed by a small volume in one corner of the room. Such extreme situations never arise naturally in systems left undisturbed, but arise from artificial operations performed on the system.

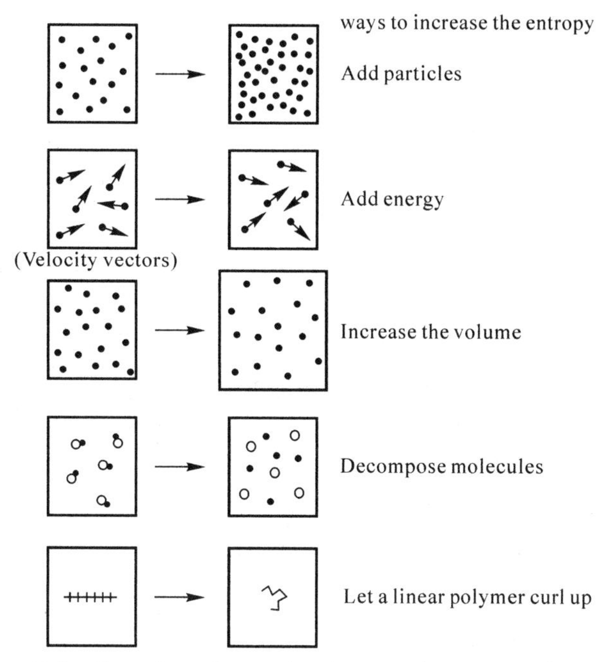

Figure 2.9 Operations that tend to increase the entropy of a system

Consider the gas in a room: the gas in one half of the room might be prepared initially with a low value of the average energy per molecule, while the gas in the other half of the room might be prepared initially with a higher value of the average energy per molecule. If the gas in the two halves is now allowed to interact by removal of a partition, the gas molecules will come very quickly to a most probable configuration in which the molecules in both halves of the room have the same average energy. Nothing else will ever be observed to happen. We will never observe the system to leave the most probable configuration and reappear later in the initial specially prepared configuration. This is true even though the equations of motion of physics are reversible in time and do not distinguish past and future.

2.8 Laws of Thermodynamics

When thermodynamics is studied as a nonstatistical subject, four postulates are introduced. These postulates are called the laws of thermodynamics. In essence, these laws are contained within our statistical formulation of thermal physics, but it is useful to exhibit them as separate statements.

Zeroth law. If two systems are in thermal equilibrium with a third system, they must be in thermal equilibrium with each other. This law is a consequence.

A large or macroscopic system may be taken to be one with more than 10′ü or 10′* atoms. The calculation of the time required for the process is largely a problem in hydrodynamics.

Of the condition Eq. (2. 20b) for equilibrium in thermal contact:

$$\left(\frac{\partial \ln g_1}{\partial U_1}\right)_{N_1} = \left(\frac{\partial \ln g_3}{\partial U_3}\right)_{N_3}, \quad \left(\frac{\partial \ln g_2}{\partial U_2}\right)_{N_2} = \left(\frac{\partial \ln g_3}{\partial U_3}\right)_{N_3}$$

In other words, $\tau_1 = \tau_3$ and $\tau_2 = \tau_3$ imply $\tau_1 = \tau_2$.

First law. Heat is a form of energy. This law is no more than a statement of the principle of conversation of energy Chapter 6 discuses what form of energy heat is.

Second law. There are many equivalent statements of the second law. We shall use the statistical statement, which we have called the law of increase of entropy, applicable when a constraint internal to a closed system is removed. The commonly used statement of the law of increase of entropy is: "If a closed system is in a configuration that is not the equilibrium configuration, the most probable consequence will be that the entropy of the system will increase monotonically in successive instants of time." This is a looser statement than the one we gave with Eq. (2. 35) above.

The traditional thermodynamic statement is the Kelvin-Planck formulation of second law of thermodynamics: "It is impossible for any cyclic process to occur whose sole effect is the extraction of heat from a reservoir and the performance of an equivalent amount of work." An engine that violates the second law by extracting the energy of one heat reservoir is said to be performing perpetual motion of the second kind. We will see that the Kelvin-Planck formulation is a consequence of the statistical statement.

Third law. The entropy of a system approaches a constant value as the temperature approaches zero. The earliest of this law, due to Nernst, is that at the absolute zero the entropy difference disappears between all those configurations of a system which are in internal thermal equilibrium. The third law follows from the statistical definition of the entropy, provided that the ground state of the system has a well—difined multiplicity. If the ground state multiplicity is $g(0)$, the corresponding entropy is $\sigma(0) = \ln g(0)$ as $\tau = 0$. From a quantum point of view, the law does not appear to say much that is not implicit in the definition of entropy, provided, however, that the system is in its lowest set of quantum states at absolute zero. Except for glasses, there would not be any objection to affirming that $g(0)$ is a small number and $\sigma(0)$ is essentially zero. Glasses have a frozen-in disorder, and for them $\sigma(0)$ can be substantial, of the order of the number of atoms N. what the third law tells us in real life is that curves of many reasonable physical quantities plotted against τ must come in flat as τ approaches 0.

2.9 Entropy as a Logarithm

Several useful properties follow from the definition of the entropy as the logarithm of the number of accessible states, instead of as the number of accessible states itself. First, the entropy of two independent systems is the sum of the separate entropies.

Second, the entropy is entirely insensitive—for all practical purposes—to the precision σU with which the energy of a closed system is defined. We have never meant to imply that the system energy is known exactly, a circumstance that for a discrete spectrum of energy eigenvalues would make the number of accessible states depend erratically on the energy. We have simply not paid much attention to the precision, whether it be determined by the uncertainty principle $\sigma U \sigma$ (time) $\sim h$, or determined otherwise. Define $D(U)$ as the number of accessible states per unit energy range; $D(U)$ can be a suitable smoothed average centered at U. Then $g(U)=D(U)\sigma(U)$ is the number of accessible states in the range σU at U. The entropy is

$$\sigma(U) = \ln D(U)\sigma U = \ln D(U) + \ln\sigma U \tag{2.36}$$

Typically as for the system of N spins, the total number of states will be of the order of 2^N. If the total energy is of the order of N times some average one particle energy Δ, then $D(U)\sim 2^N/N\Delta$. thus

$$\sigma(U) = N\ln 2 - \ln N\Delta + \ln\sigma U \tag{2.37}$$

Let $N=10^{20}$, $\Delta=10^{-14}$ erg, and $\sigma U=10^{-1}$ erg,

$$\sigma(U) = 0.69\times 10^{20} - 13.82 - 2.3 \tag{2.38}$$

We see from this example that the value of the entropy is dominated overwhelmingly by the value of N; the precision σU is without perceptible effect on the result. In the problem of N free particles in a box, the number of states is proportional to something like $U^N\sigma U$, when $\sigma\sim N\ln U+\ln\sigma U$. Again the term in N is dominant, a conclusion independent of even thy system of units used for the energy.

Example: Perpetual motion of the second kind

Early in our study of physics we came to understand the impossibility of a perpetual motion machine, a machine that will give forth more energy than it absorbs.

Equally impossible is a perpetual motion machine of the second kind, as it is called, in which heat is extracted from part of the environment and delivered to another part of the environment, the difference in temperature thus established being used to power a heat engine that delivers mechanical work available for any purpose at no cost to us. In brief, we cannot propel a ship. The spontaneous transfer of energy from the low temperature ocean to a higher temperature boiler on the ship would decrease the total entropy of the combined systems and would thus be in violation of the law of increase of entropy.

2.10 Summary

(1)The fundamental assumption is that a closed system is equally likely to be in any of the quantum states accessible to it.

(2)If $P(s)$ is the probability that a system is in the states, the average value of a quantity X is

$$\langle X\rangle = \sum_s X(s)P(s)$$

(3)An ensemble of systems is composed of very many systems, all constructed alike.

(4)The number of accessible states of the combined systems 1 and 2 is

$$g(s) = \sum_s g_1(s_1)g_2(s-s_1)$$

where $s_1+s_2=s$.

(5)The entropy $\sigma(N,U)\equiv\ln g(N,U)$. The relation $S=k_B\sigma$ connects the conventional entropy σ.

(6)The fundamental temperature $\tau=k_B T$ connects the fundamental temperature and the conventional temperature.

(7)The law of increase of entropy states that the entropy of a closed system tends to remain constant or to increase when a constant or to increase when a constraint internal to the system is removed.

(8)The thermal equilibrium values of then physical properties of a system are defined as averages over all states accessible when the system is in contact with a large system or reservoir. If the first system also is large, the thermal equilibrium properties are given accurately by consideration of the states in the most probable configuration alone.

2.11 Problems

1. Entropy and temperature

Suppose $g(U)=CU^{3N/2}$, where C is a constant and N is the number of particles.

(1)Show that $U=\frac{3}{2}N\tau$.

(2)Show that $(\partial^2\sigma/\partial U^2)_N$ is negative. This form of $g(U)$ actually applies to an ideal gas.

2. Paramagnetism

Find the equilibrium value at temperature τ of the fractional magnetization

$$M/Nm = 2\langle s\rangle/N$$

Of the system of N spins each of magnetic moment m in a magnetic field B. The spin excess is 2 s. Take the entropy as the logarthithm of the multiplicity $g(N,s)$ as given in Eq. (1.34):

$$\sigma(s) \approx \ln g(N,0) - 2s^2/N \tag{2.39}$$

For $|s| \ll N$. Further, Show that in this approximation

$$\sigma(U) = \sigma_0 - U^2/2m^2B^2N \tag{2.40}$$

With $\sigma_0 = \ln g(N,0)$. Further that $1/\tau = -U/m^2B^2N$, where U denotes $\langle U\rangle$, the thermal average energy.

3. Quantum harmonic oscillator

(1)Find the entropy of a set of N oscillators of frequency ω as a function of the total quantum number n. Use the multiplicity function of Eq. (1.54) and make the Stirling approximation $\ln N! \approx N\ln N - N$. Replace $N-1$ by N.

(2)Let U denote the total energy $nh\omega$ of the total energy at temperature τ is

$$U = \frac{Nh\omega}{\exp(h\omega/\tau) - 1} \tag{2.41}$$

This is the Planck result; it is derived again in Chapter 4 by a powerful method that does not require us to find the multiplicity function.

4. The meaning of "never"

It has been said that "six monkeys, set to strum unintelligently on typewriters for millions of years, would be bound in time to write all the books in the British Museum." This statement is nonsense, for it gives a misleading conclusion about very, very large numbers. Could all the monkeys in the world have typed out a single specified book in the age of the universe?

Suppose that 10^{10} monkeys have been seated at typewrites throughout the age of the universe, 10^{18} s. This number of monkeys is about three times greater than the present human population of the earth. We suppose that a monkey can hit 10 typewriter keys per second. A typewriter may have 44 keys; we accept lowercase letters in place of capital letters. Assuming that Shakespeare's Hamlet has 10^5 characters, will the monkeys hit upon Hamlet?

(1)Show that the probability that any given sequence of 10^5 characters typed at random will come out in the correct sequence(the sequence of Hamlet) is of the order of

$$\left(\frac{1}{44}\right)^{100,000}=10^{-164,345}$$

where we have used $\lg 44=1.643,45$.

(2)Show that the probability that a monkey-Hamlet will be typed in the age of the Universe is approximately$10^{-164,345}$. The probability of Hamlet is therefore zero in any operational sense of an event, so that the original statement at the beginning of this problem is nonsense: one book, much less a library, will never occur in the total literary production of the monkeys.

5. Acditivity of entropy for two spin systems

Given two systems of $N_1 \approx N_2=10^{22}$ spins with multiplicity functions $g_1(N_1,s_1)$ and $g_2(N_2,s-s_1)$, the product g_1g_2 as a function of s_1 is relatively sharply peaked at $s_1=\hat{s}_1$. For $s_1=\hat{s}_1+10^{12}$, the product g_1g_2 is reduced by 10^{-174} from its peak value. Use the Gaussian approximation to the multiplicity function; the form Eq. (2. 17) may be useful.

(1)Compute $g_1g_2/(g_1g_2)_{\max}$ for $s_1=\hat{s}_1+10^{11}$ and $s=0$.

(2) For $s_1=10^{20}$, by what factor must you multiply $(g_1g_2)_{\max}$ to make it equal to $\sum_{s_1} g_1(N_1,s_1)g_2(N_1,s-s_1)$; give the factor to the nearest order of magnitude.

(3)How large is the fractional error in the entropy when you ignore this factor?

6. Integrated deviation

For the example that gave the result of Eq. (2. 17), calculate approximately the probability that fractional deviation from equilibrium σ/N_1 is 10^{-10} or larger. Take $N_1=N_2=10^{22}$. You will find it convenient to use an asymptotic expansion for the complementary error function. When $x\geqslant 1$,

$$2x\exp(x^2)\int_x^{\infty}\exp(-t^2)\mathrm{d}t \approx 1+\text{small terms}$$

Chapter 3 Boltzmann Distribution and Helmholtz Free Energy

In this chapter we develop the principles that permit us to calculate the values of the physical properties of a system as a function of the temperature. We assume that the system S of interest to us is in thermal equilibrium with a very large system R, called the reservoir. The system and the reservoir will have a common temperature τ because they are in thermal contact.

The total system R+S is a closed system, insulated from all extemal influences, as in Figure 3. 1. The total energy $U_0 = U_R + U_S$ is constant. In particular, if the system is in a state of energy ε_s, then $U_0 - \varepsilon_s$ is the energy of the reservoir.

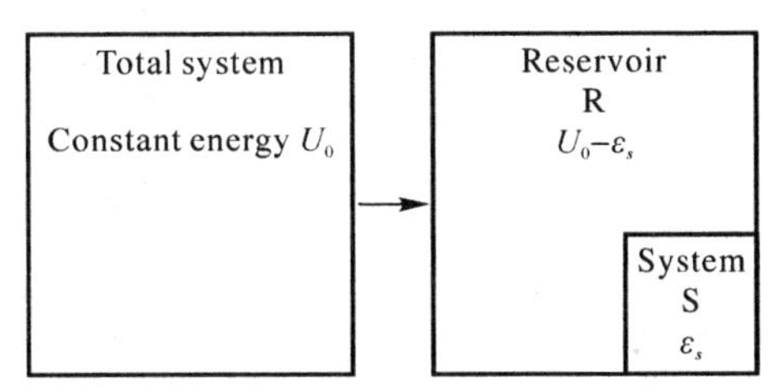

Figure 3. 1 Representation of a closed total system decomposed into a reservoir R in thermal contact with a system S

3. 1 Boltzmann Factor

A central problem of thermal physics is to find the probability that the system S will be in a specific quantum state s of energy ε_s. This probability is proportional to the Boltzmann factor.

When we specify that S should be in the state s, the number of accessible states of the total system is reduced to the number of accessible states of the reservoir R, at the appropriate energy. That is, the number g_{R+S} of states

$$R+S = g_R \tag{3. 1}$$

because for our present purposes we have specified the state of S.

If the system energy is ε_s, the reservoir energy is $U_0 - \varepsilon_s$ as in Figure 3. 2. The ratio of the probability that the system is in quantum state 1 at energy ε_1 to the probability that the system is in quantum state 2 at energy ε_2 is the ratio of the two multiplicities:

$$\frac{P(\varepsilon_1)}{P(\varepsilon_2)}=\frac{\text{Multiplicity of R at energy } U_0-\varepsilon_1}{\text{Multiplicity of R at energy } U_0-\varepsilon_2}=\frac{g_R(U_0-\varepsilon_1)}{g_R(U_0-\varepsilon_2)} \tag{3.2}$$

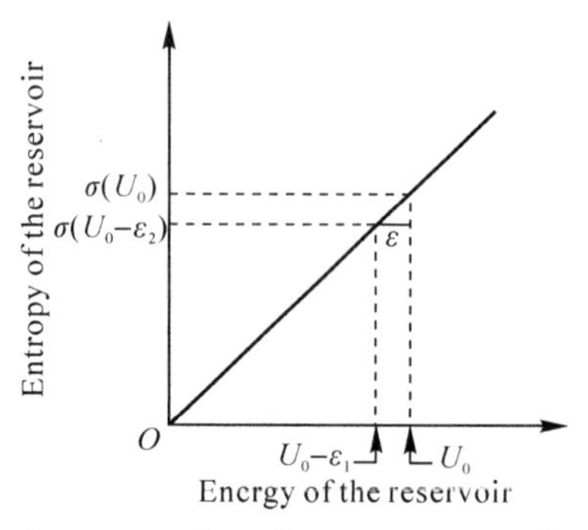

Figure 3.2 The change of entropy when the reservoir transfers energy ε to the system. The fractional effect of the transfer on the reservoir is small when the reservoir is large, because a large reservoir will have a high entropy

This result is a direct consequence of what we have called the fundamental assumption. The two situations are shown in Figure 3.3. Although questions about the system depend on the constitution of the reservoir, we shall see that the dependence is only on the temperature of the reservoir.

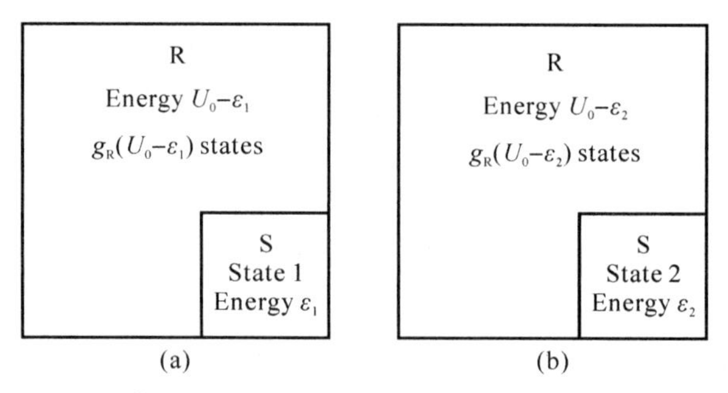

Figure 3.3 The system in(a), (b)is in quantum state 1,2. The reservoir has $g_R(U_0-\varepsilon_1)$, $g_R(U_0-\varepsilon_2)$ accessible quantum states, in (a)and (b) respectively

If the reservoirs are very large. the multiplicities are very, very large numbers. We write Eq. (3.2) in terms of the entropy of the reservoir:

$$\frac{P(\varepsilon_1)}{P(\varepsilon_2)}=\frac{\exp[\sigma_R(U_0-\varepsilon_1)]}{\exp[\sigma_R(U_0-\varepsilon_2)]}=\exp[\sigma_R(U_0-\varepsilon_1)-\sigma_R(U_0-\varepsilon_2)] \tag{3.3}$$

with

$$\Delta\sigma_R\equiv\sigma_R(U_0-\varepsilon_1)-\sigma_R(U_0-\varepsilon_2) \tag{3.4}$$

the probability ratio for the two states 1, 2 of the system is simply

$$\frac{P(\varepsilon_1)}{P(\varepsilon_2)} = \exp(\Delta\sigma_R) \tag{3.5}$$

Let us expand the entropies in Eq. (3.4) in a Taylor series expansion about $\sigma_R(U_0)$. The Taylor series expansion of $f(x)$ about $f(x_0)$ is

$$f(x_0 + a) = f(x_0) + a\left(\frac{\mathrm{d}f}{\mathrm{d}x}\right)_{x=x_0} + \frac{1}{2!}a^2\left(\frac{\mathrm{d}^2 f}{\mathrm{d}x^2}\right)_{x=x_0} + \cdots \tag{3.6}$$

Thus

$$\sigma(U_0 - \varepsilon) = \sigma_R(U_0) - \varepsilon\left(\frac{\partial\sigma_R}{\partial U}\right)_{V,N} + \cdots = \sigma_R(U_0) - \frac{\varepsilon}{\tau} + \cdots \tag{3.7}$$

where $1/\tau \equiv \left(\frac{\partial\sigma_R}{\partial U}\right)_{V,N}$ gives the temperature. The partial derivative is taken at energy U_0.

The higher order terms in the expansion vanish in the limit of an infinitely large reservoir.

Therefore $\Delta\sigma_R$ defined by Eq. (3.4) becomes

$$\Delta\sigma_R = -(\varepsilon_1 - \varepsilon_2)/\tau \tag{3.8}$$

The final result of Eq. (3.5) and Eq. (3.8) is

$$\frac{P(\varepsilon_1)}{P(\varepsilon_2)} = \frac{\exp\left(-\frac{\varepsilon_1}{\tau}\right)}{\exp\left(-\frac{\varepsilon_2}{\tau}\right)} \tag{3.9}$$

A term of the form $\exp(-\varepsilon/\tau)$ is known as a Boltzmann factor. This result is of vast utility. It gives the ratio of the probability of finding the system in a single quantum state 1 to the probability of finding the system in a single quantum state 2.

3.2 Partition Function

It is helpful to consider the function

$$Z(\tau) = \sum_s \exp(-\varepsilon_s/\tau) \tag{3.10}$$

called the partition function. The summation is over the Boltzmann factor $\exp(-\varepsilon_s/\tau)$ for all states s of the system. The partition function is the pro-portionality factor between the probability $P(\varepsilon_s)$ and the Boltzmann factor $\exp(-\varepsilon_s/\tau)$:

$$P(\varepsilon_s) = \frac{\exp(-\varepsilon_s/\tau)}{Z} \tag{3.11}$$

We see that $\sum P(\varepsilon_s) = \frac{Z}{Z} = 1$: the sum of all probabilities is unity.

The result of Eq. (3.11) is one of the most useful results of statistical physics. The average energy of the system is $U = \langle \varepsilon \rangle = \sum \varepsilon_s P(\varepsilon_s)$, or

$$U = \frac{\sum \varepsilon_s \exp(-\varepsilon_s/\tau)}{Z} = \tau^2 (\partial \ln Z / \partial \tau) \tag{3.12}$$

We expand $\sigma(U_0 - \varepsilon)$ and not $g(U_0 - \varepsilon)$ because the expansion of the latter quantity immediately gives convergence difficulties.

The average energy refers to those states of a system that can exchange energy with a reservoir. The notation $\langle \cdots \rangle$ denotes such an average value and is called the thermal average or ensemble average. In Eq. (3.12) the symbol U is used for $\langle \varepsilon \rangle$ in conformity with common practice; U will now refer to the system and not, as earlier, to the system+reservoir.

Example: Energy and heat capacity of a two state system

We treat a system of one particle with two states, one of energy 0 and one of energy ε. The particle is in thermal contact with a reservoir at temperature τ. We want to find the energy and the heat capacity of the system as a function of the temperature τ. The partition function for the two states of the particle is

$$Z = \exp\left(-\frac{0}{\tau}\right) + \exp\left(-\frac{\varepsilon}{\tau}\right) = 1 + \exp\left(-\frac{\varepsilon}{\tau}\right) \tag{3.13}$$

The average energy is

$$U \equiv \langle \varepsilon \rangle = \frac{\varepsilon \exp\left(-\frac{\varepsilon}{\tau}\right)}{Z} = \varepsilon \frac{\exp\left(-\frac{\varepsilon}{\tau}\right)}{1 + \exp\left(-\frac{\varepsilon}{\tau}\right)} \tag{3.14}$$

This function is plotted in Figure 3.4.

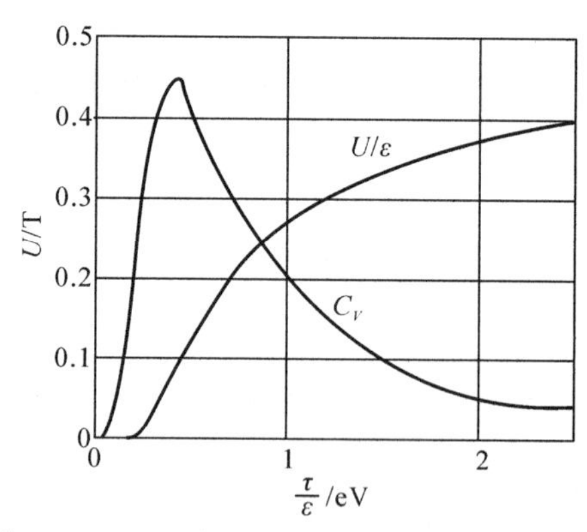

Figure 3.4 Energy and heat capacity of a two state system as functions of the temperature τ. The energy is plotted in units of ε

If we shift the zero of energy and take the energies of the two states as $-\frac{1}{2}\varepsilon$ and $+\frac{1}{2}\varepsilon$, instead of as 0 and ε, the results appear differently. We have

$$Z = \exp(\varepsilon/2\tau) + \exp(-\varepsilon/2\tau) = 2\cosh(\varepsilon/2\tau) \tag{3.15}$$

and

$$\langle\varepsilon\rangle = \frac{\left(-\frac{1}{2}\varepsilon\right)\exp\left(\frac{\varepsilon}{2\tau}\right)+\left(\frac{1}{2}\varepsilon\right)\exp\left(-\frac{\varepsilon}{2\tau}\right)}{Z} = -\varepsilon\frac{\sinh\left(\frac{\varepsilon}{2\tau}\right)}{2\cosh\left(\frac{\varepsilon}{2\tau}\right)} = -\frac{1}{2}\varepsilon\tanh\left(\frac{\varepsilon}{2\tau}\right) \tag{3.16}$$

The heat capacity C_V of a system at constant volume is defined as

$$C_V \equiv \tau(\partial\sigma/\partial\tau)_V \tag{3.17a}$$

which by the thermodynamic identity Eq. (3.34a) derived below is equivalent to the alternate definition:

$$C_V \equiv (\partial U/\partial\tau)_V \tag{3.17b}$$

We hold V constant because the values of the energy are calculated for a system at a specified volume. From Eq. (3.14) and Eq. (3.17b),

$$C_V = \varepsilon\frac{\partial}{\partial\tau}\frac{1}{\exp\left(\frac{\varepsilon}{\tau}\right)+1} = \left(\frac{\varepsilon}{\tau}\right)^2\frac{\exp\left(\frac{\varepsilon}{\tau}\right)}{\left[\exp\left(\frac{\varepsilon}{\tau}\right)+1\right]^2} \tag{3.18a}$$

The same result follows from Eq. (3.16).

In conventional units C_V is defined as $T(\partial S/\partial T)_V$ or $(\partial U/\partial T)_V$, whence (conventional)

$$C_V = k_B\left(\frac{\varepsilon}{k_B T}\right)^2\frac{\exp(\varepsilon/k_B T)}{\left[\exp\left(\frac{\varepsilon}{k_B T}\right)+1\right]^2} \tag{3.18b}$$

In fundamental units the heat capacity is dimensionless; in conventional units it has the dimensions of energy per Kelvin. The specific heat is defined as the heat capacity per unit mass.

The hump in the plot of heat capacity versus temperature in Figure 3.4 is called a Schottky anomaly. For $\tau \gg \varepsilon$ the heat capacity Eq. (3.18a) becomes

$$C_V \approx (\varepsilon/2\tau)^2 \tag{3.19}$$

Notice that $C_V \propto \tau^{-2}$ in this high temperature limit. In the low temperature limit the temperature is small in comparison with the energy level spacing ε. For $\tau \ll \varepsilon$ we have

$$C_V \approx \left(\frac{\varepsilon}{\tau}\right)^2\exp(-\varepsilon/\tau) \tag{3.20}$$

The exponential factor $\exp(-\varepsilon/\tau)$ reduces C_V rapidly as τ decreases, because $\exp(-1/x) \to 0$ as $x \to 0$.

Definition: Reversible process. A process is reversible if carried out in such a way that the system is always infinitesimally close to the equilibrium condition. For example, if the entropy is a function of the volume, any change of volume must be carried out so slowly that the entropy at any volume V is closely equal to the equilibrium entropy $\sigma(V)$. Thus, entropy is well defined at every stage of a reversible process, and by reversing the direction of the change the system will be returned to its initial condition. In reversible processes, the condition of the system is well defined at all times, in contrast to irreversible processes, where usually we will not know what is going on during the process. We cannot apply the mathematical methods of thermal physics to systems whose condition is undefined.

A volume change that leaves the system in the same quantum state is an example of an isentropic reversible process. If the system always remains in the same state the entropy change will be zero between any two stages of the process, because the number of states in an ensemble of similar system does not change. Any process in which the entropy change vanishes is an isentropic reversible process, But reversible processes are not limited to isentropic processes, and we shall have a special interest also in isothermal reversible processes.

3.3 Pressure

Consider a system in the quantum state s of energy ε_s. We assume ε_s to be a function of the volume of the system. The volume is decreased slowly from V to $V-\Delta V$ by application of an external force. Let the volume change take place sufficiently slowly that the system remains in the same quantum state s throughout the compression. The "same" state may be characterized by its quantum numbers (see Figure 3.5) or by the number of zeros in the wave function.

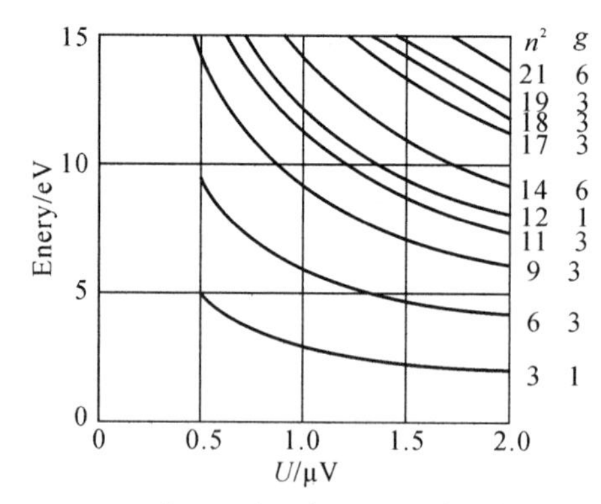

Figure 3.5 Dependence of energy on volume, for the energy levels of a free particle confined to a cube

The energy of the states after the reversible volume change is

$$\varepsilon_s(V-\Delta V)=\varepsilon_s(V)-\left(\frac{d\varepsilon_s}{dV}\right)\Delta V+\cdots \tag{3.21}$$

Consider a pressure p_s, applied normal to all faces of a cube. The mechanical work done on the system by the pressure in a contraction (see Figure 3.6) of the cube volume from V to $V-\Delta V$ appears as the change of energy of the s

$$U(V-\Delta V)-U(V)=\Delta V=-\left(\frac{\mathrm{d}\varepsilon_s}{\mathrm{d}V}\right)\Delta V \tag{3.22}$$

The curves are labeled by $n^2=n_x{}^2+n_y{}^2+n_z{}^2$, as in Figure 1.2. The multiplicities g are also given. The volume change here is isotropic: a cube remains a cube. The energy range δ_ε of the states represented in an ensemble of systems will increase in a reversible compression, but we know from the discussion in Chapter 2 that the width of the energy range itself is of no practical importance. It is the change in the average energy that is important.

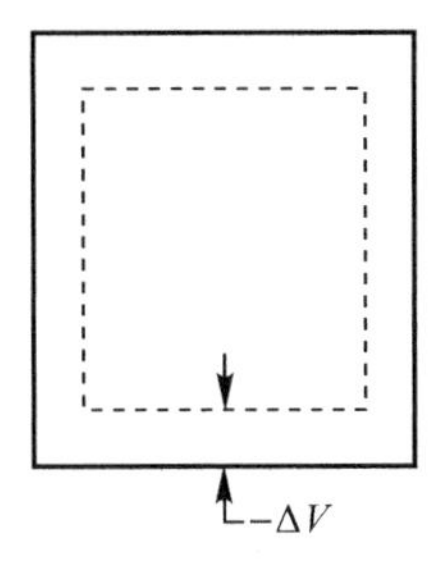

Figure 3.6 Volume change $-\Delta V$ in uniform compression of a cube

Here U denotes the energy of the system. Let A be the area of one face of the cube; then

$$A(\Delta x+\Delta y+\Delta z)=\Delta V \tag{3.23}$$

If all increments ΔV and $\Delta x=\Delta y=\Delta z$ are taken as positive in the compression. The work done in the compression is

$$\Delta U=P_s A(\Delta x+\Delta y+\Delta z)=P_s\Delta V \tag{3.24}$$

so that, on comparison with Eq. (3.22),

$$P_s=-\mathrm{d}\varepsilon_s/\mathrm{d}V \tag{3.25}$$

is the pressure on a system in the state s.

We average Eq. (3.25) over all states of the ensemble to obtain the average pressure $\langle P\rangle$, usually written as p:

$$p=-\left(\frac{\partial U}{\partial V}\right)_\sigma \tag{3.26}$$

where $U\equiv\langle\varepsilon\rangle$. The entropy σ is held constant in the derivative because the number of states in the ensemble is unchanged in the reversible compression we have described. We have a collection of systems, each in some state, and each remains in this state in the compression.

The result of Eq. (3.26) corresponds to our mechanical picture of the pressure on a system that is maintained in some specific state. For applications we shall need also the later result of Eq. (3.50) for the pressure on a system maintained at constant temperature.

We look for other expressions for the pressure. The number of states and thus the entropy depend only on U and on V, for a fixed number of particles, so that only the two variables U and V describe the system. The differential of the entropy is

$$\mathrm{d}\sigma(U,V) = \left(\frac{\partial\sigma}{\partial U}\right)_V \mathrm{d}U + \left(\frac{\partial\sigma}{\partial V}\right)_U \mathrm{d}V \tag{3.27}$$

This gives the differential change of the entropy for arbitrary independent differential changes $\mathrm{d}U$ and $\mathrm{d}V$. Assume now that we select $\mathrm{d}U$ and $\mathrm{d}V$ inter-dependently, in such a way that the two terms on the right-hand side of Eq. (3.27) cancel. The overall entropy change $\mathrm{d}\sigma$ will be zero. If we denote these inter-dependent values of $\mathrm{d}U$ and $\mathrm{d}V$ by $(\delta U)_\sigma$ and $(\delta V)_\sigma$, the entropy change will be zero:

$$0 = \left(\frac{\partial\sigma}{\partial U}\right)_V (\delta U)_\sigma + \left(\frac{\partial\sigma}{\partial V}\right)_U (\delta V)_\sigma \tag{3.28}$$

After division by $(\delta V)_\sigma$,

$$0 = \left(\frac{\partial\sigma}{\partial U}\right)_V \frac{(\delta U)_\sigma}{(\delta V)_\sigma} + \left(\frac{\partial\sigma}{\partial V}\right)_U \tag{3.29}$$

But the ratio $\frac{(\delta U)_\sigma}{(\delta V)_\sigma}$ is the partial derivative of U with respect to V at constant σ:

$$\frac{(\delta U)_\sigma}{(\delta V)_\sigma} \equiv \left(\frac{\partial U}{\partial V}\right)_\sigma \tag{3.30}$$

With this and the definition $1/\tau \equiv (\partial\sigma/\partial U)_V$, Eq. (3.29) becomes

$$\left(\frac{\partial U}{\partial V}\right)_\sigma = -\tau\left(\frac{\partial\sigma}{\partial V}\right)_U \tag{3.31}$$

By Eq. (3.26) the left-hand side of Eq. (3.31) is equal to $-p$, whence

$$p = \tau\left(\frac{\partial\sigma}{\partial V}\right)_U \tag{3.32}$$

3.4 Thermodynamic Identity

Consider again the differential Eq. (3.27) of the entropy; substitute the new result for the pressure and the definition of τ to obtain

$$\mathrm{d}\sigma = \frac{1}{\tau}\mathrm{d}U + \frac{p}{\tau}\mathrm{d}V \tag{3.33}$$

or

$$\tau\mathrm{d}\sigma = \mathrm{d}U + p\mathrm{d}V \tag{3.34a}$$

This useful relation will be called the thermodynamic identiy. The form with N variable will appear in Eq. (5.38). A simple transposition gives

$$dU = \tau d\sigma - p dV \text{ or } dU = T dS - p dV \tag{3.34b}$$

If the actual process of change of state of the system is reversible, we can identify $\tau d\sigma$ as the heat added to the system and $-pdV$ as the work done on the system. The increase of energy ia caused in part by mechanical work and in part by the transfer of heat. Heat is defined as the transfer of energy between two systems brought into thermal contact.

3.5 Helmholtz Free Energy

The function

$$F \equiv U - \tau\sigma \tag{3.35}$$

is called the Helmholtz free energy. This function plays the part in thermal physics at constant temperature that the energy U plays in ordinary mechanical processes, which are always understood to be at constant entropy, because no internal changes of state are allowed. The free energy tells us how to balance the conflicting demands of a system for minimum energy and maximum entropy. The Helmholtz free energy will be a minimum for a system R in thermal contact with a reservoir R, if the volume of the system is constant.

We first show that F is an extremum in equilibrium at constant τ and V. By definition, for infinitesimal reversible transfer from R to S,

$$dU_S = T dS - d p dV \tag{3.36}$$

$$dF = 0 \tag{3.37}$$

which is the condition for F to be an extremum with respect to all variations at constant volume and temperature. We like F because we can calculate it from the energy eigenvalues of the system. comment. We can show that the extremum is a minimum. The total energy is $U=U_S+U_R$. Then the entropy is

$$U = U_S + U_R \tag{3.38}$$

We know that

$$\frac{(\delta U)_\sigma}{(\delta V)_\sigma} \equiv \left(\frac{\partial U}{\partial V}\right)_\sigma \tag{3.39}$$

so that Eq. (3.38) becomes

$$-2smB + \left(\frac{1}{2}N + s\right)\tau \ln\left(\frac{1}{2} + \frac{s}{N}\right) \tag{3.40}$$

The free energy of the system at constant τ, V will increase for any departure from the equilibrium configuration.

Example: Minimum property of the free energy of a paramagnetic system

Consider the model system of Chapter 1, with N_1 spins up and N_1 spins down. Let $N = N_1 + N_1$; the spin excess is $2s = N_1 - N_1$. The entropy in the Stirling approximation is found with the help of an approximate form of Eq. (1.30):

$$\sigma(s) \approx -\left(\frac{1}{2}N + s\right)\ln\left(\frac{1}{2} + \frac{s}{N}\right) - \left(\frac{1}{2}N - s\right)\ln\left(\frac{1}{2} - \frac{s}{N}\right) \tag{3.41}$$

The energy in a magnetic field B is $-2smB$, where m is the magnetic moment of an elementary magnet. The free energy function is $F_L(\tau, s, B) \equiv U(s, B) - \tau\sigma(s)$, or

$$F_L(\tau, s, B) = -2smB + \left(\frac{1}{2}N + s\right)\tau \ln\left(\frac{1}{2} + \frac{s}{N}\right) + \left(\frac{1}{2}N - s\right)\tau \ln\left(\frac{1}{2} - \frac{s}{N}\right) \tag{3.42}$$

At the minimum of $F_L(\tau, s, B)$ with respect to s, this function becomes equal to the equilibrium minimum free energy $F(\tau, B)$. That is, $F_L(\tau, <s>, B) = F(\tau, B)$, because $<s>$ is a function of τ and B. The minimum of F_L with respect to the spin excess occurs when

$$(\partial F_L / \partial s)_{\tau, B} = 0 = -2mB + \tau \ln \frac{N + 2s}{N - 2s} \tag{3.43}$$

Thus in the magnetic field B the thermal equilibrium value of the spin excess $2s$ is given by

$$\frac{N + <2s>}{N - <2s>} = \exp\left(\frac{2mB}{\tau}\right), \quad <2s> = N\left[\frac{\exp\left(\frac{2mB}{\tau}\right) - 1}{\exp\left(\frac{2mB}{\tau}\right) + 1}\right] \tag{3.44}$$

Or, on dividing numerator and denominator by $\exp\left(\frac{mB}{\tau}\right)$,

$$<2s> = N\tanh(mB/\tau) \tag{3.45}$$

The magnetization M is the magnetic moment per unit volume. If n is the number of spins per unit volume, the magnetization in themal equlibrium in the magnetic field is

$$M = <2s> \frac{m}{V} = nm\tanh(mB/\tau) \tag{3.46}$$

The free energy of the system in equilibrium can be obtained by substituting Eq. (3.45) in Eq. (3.42). It is easier, however, to obtain F directly from the partition function for one magnet:

$$Z = \exp\left(\frac{mB}{\tau}\right) + \exp\left(-\frac{mB}{\tau}\right) = 2\cosh(mB/\tau) \tag{3.47}$$

Now use the relation $F = -\tau \ln Z$ as derived below. Multiply by N to obtain the result for N magnets.

3.6 Differential Relations

The differential of F is

$$dF = dU - \tau d\sigma - \sigma d\tau$$

or, with use of the thermodynamic identity (3.34a),

$$dF = -\sigma d\tau - p dV \tag{3.48}$$

for which

$$\left(\frac{\partial F}{\partial \tau}\right)_V = -\sigma, \quad \left(\frac{\partial F}{\partial V}\right)_\tau = -p \tag{3.49}$$

These relations are widely used.

The free energy F in the result $p = -\left(\frac{\partial F}{\partial V}\right)_\tau$, acts as the effective energy for an isothermal change of volume: contrast this result with Eq. (3.26). The result may be written as

$$p = -\left(\frac{\partial U}{\partial V}\right)_\tau + \tau\left(\frac{\partial \sigma}{\partial V}\right)_\tau \tag{3.50}$$

by use of $F \equiv U - \tau\sigma$. The two terms on the right-hand side of Eq. (3.50) represent what we may call the energy pressure and the entropy pressure. The energy pressure $-(\partial U/\partial V)_\tau$ is dominant in most solids and the entropy pressure $\tau(\partial\sigma/\partial V)_\tau$ is dominant in gases and in elastic polymers such as rubber. The entropy contribution is testimony of the importance of the entropy: the naive feeling from simple mechanics that $-dU/dV$ must tell everything about the pressure is seriously incomplete for a process at constant temperature, because the entropy can change in response to the volume change even if the energy is independent of volume, as for an ideal gas at constant temperature.

Maxwell relation. We can now derive one of a group of usefulthermodynamic relations called Maxwell relations. Form the cross-derivatives $\partial^2 F/\partial V\partial\tau$ and $\partial^2 F/\partial\tau\partial V$, which must be equal to each other. It follows from Eq. (3.49) that

$$(\partial\sigma/\partial V)_\tau = (\partial p/\partial\tau)_V \tag{3.51}$$

a relation that is not at all obvious. Other Maxwell relations will be derived later at appropriate points, by similar arguments. The methodology of obtaining thermody—namic relations is discussed by R. Gilmore, J. Chem. Phys. 75, 5964 (1981).

Calculation of F from Z. Becasue $F \equiv U - \tau\sigma$ and $\sigma = -(\partial F/\partial\tau)_V$, we have the diferential equation:

$$F = U + \tau(\partial F/\partial\tau)_V \quad \text{or} \quad -\frac{\tau^2 \partial\left(\frac{F}{\tau}\right)}{\partial\tau} = U \tag{3.52}$$

We show that this equation is satisfied by

$$\frac{F}{\tau} = -\ln Z \tag{3.53}$$

where Z is the partition function. On substitution,

$$\frac{\partial\left(\frac{F}{\tau}\right)}{\partial\tau} = -\frac{\partial \ln Z}{\partial \tau} = -U/\tau^2 \tag{3.54}$$

by Eq. (3.12). This proves that

$$F = -\tau \ln Z \tag{3.55}$$

satisfies the required differential Eq. (3.52).

It would appear possible for F/τ to contain an additive constant α such that $F = -\tau\ln Z + \alpha\tau$. However, the entropy must reduce to $\ln g_0$ when the temperature is so low that only the g_0 coincident states at the lowest energy ε_0 are occupied. In that limit $\ln Z \to \ln g_0 - \varepsilon_0/\tau$, so that $\sigma = -\frac{\partial F}{\partial \tau} \to \frac{\partial(\tau \ln Z)}{\partial \tau} = \ln g_0$ only if $\alpha = 0$.

We may write the result as

$$Z = \exp(-F/\tau) \tag{3.56}$$

And the Boltzamann factor Eq. (3.11) for the occupancy probability of a quantum state s becomes

$$P(\varepsilon_s) = \frac{\exp(-\varepsilon_s/\tau)}{Z} = \exp[(F - \varepsilon_s)/\tau] \tag{3.57}$$

3.7 Ideal Gas: a First Look

One atom in a box. We calculate the partition function Z_1 of one atom of mass M free to move in a cubical box of volume $V = L^3$. The orbitals of the free particle wave equation $-\left(\frac{\hbar^2}{2M}\right)\nabla^2\psi = \varepsilon\psi$ are

$$(x, y, z) = A\sin(n_x\pi x/L)\sin(n_y\pi y/L)\sin(n_z\pi z/L) \tag{3.58}$$

where n_x, n_y, n_z are any positive integers, as in Chaper 1. Negative integers do not give independent orbitals, and a zero does not give a solution. The energy values are

$$\varepsilon_n = \frac{\hbar^2}{2M}\left(\frac{\pi}{L}\right)^2 (n_x{}^2 + n_y{}^2 + n_z{}^2) \tag{3.59}$$

We neglect the spin and all other structure of the atom, so that a state of the system is entirely specified by the values of n_x, n_y, n_z.

The partition function is the sum over the states of Eq. (3.59):

$$Z_1 = \sum_{n_x}\sum_{n_y}\sum_{n_z}\exp[-\hbar^2\pi^2(n_x^{\ 2}+n_y^{\ 2}+n_z^{\ 2})/2ML^2\tau] \tag{3.60}$$

Provided the spacing of adjacent energy values is small in comparison with τ, we may replace the summations by integrations:

$$Z_1 = \int_0^\infty \mathrm{d}n_x \int_0^\infty \mathrm{d}n_y \int_0^\infty \mathrm{d}n_z \exp[-\alpha^2(n_x^{\ 2}+n_y^{\ 2}+n_z^{\ 2})] \tag{3.61}$$

The notation $\alpha^2 \equiv \hbar^2\ \pi^2/2ML^2\tau$ is introduced for convenience. The exponential may be written as the product of three factors:

$$\exp(-\alpha^2\ n_x^{\ 2})\exp(-\alpha^2\ n_y^{\ 2})\exp(-\alpha^2\ n_z^{\ 2})$$

so that

$$Z_1 = \left[\int_0^\infty \mathrm{d}n_x \exp(-\alpha^2\ n_x^{\ 2})\right]^3 = (1/\alpha)^3\left[\int_0^\infty \mathrm{d}x\exp\ (-x^2)\right]^3 = \pi^{3/2}/8\alpha^3$$

whence

$$Z_1 = \frac{V}{(2\pi\hbar^2/M\tau)^{3/2}} = n_Q V = n_Q/n \tag{3.62}$$

in terms of the concentration $n=1/V$.

Here

$$n_Q \equiv (M\tau/2\pi\hbar^2)^{3/2} \tag{3.63}$$

is called the quantum concentration. It is concentration associated with one atom in acube of side equal to the thermal average de Broglie wavelength, which is a length roughly equal to $\frac{\hbar}{M}\langle v\rangle \sim \hbar/(M\tau)^{1/2}$. Here $\langle v\rangle$ is a thermal average velocity. This concentration will keep turning up in the thermal physics of gases, in semiconductor theory, and in the theory of chemical reactions.

For helium at atmospheric pressure at room temperature, $n \approx 2.5\times10^{19}\ \mathrm{cm}^{-3}$ and $n_Q \approx 0.8\times10^{25}\ \mathrm{cm}^{-3}$. Thus, $n/n_Q \approx 3\times10^{-6}$, which is very small compared to unity, so that helium is very dilute under normal conditions. Whenever $n/n_Q \ll 1$ we say that the gas is in the classical regime. An ideal gas is defined as a gas of noninteracting atoms in the classical regime.

The thermal average energy of the atom in the box is, as in Eq. (3.12).

$$U = \frac{\sum_n \varepsilon_n \exp\ (-\varepsilon_n/\tau)}{Z_1} = \tau^2(\partial \ln Z_1/\partial\tau) \tag{3.64}$$

because $Z_1^{\ -1}\exp\ (-\varepsilon_n/\tau)$ is the probability the system is in the state n. From Eq. (3.62), $\ln Z_1 = -\frac{3}{2}\ln(1/\tau) +$ terms independent of τ. So that for an ideal gas of one atom

$$U = \frac{3}{2}\tau \tag{3.65}$$

If $\tau = k_B T$, where k_B is the Boltzmann constant, then $U = \frac{3}{2} k_B T$, the well-known result for the energy per atom of an ideal gas.

The thermal average occupancy of a free particle orbital satisfies the in-equality:

$$Z_1^{-1} \exp\left(-\frac{\varepsilon_n}{\tau}\right) < Z_1^{-1} = n/n_Q$$

which sets an upper limit of 4×10^{-6} for the occupancy of an orbital by a helium atom at standard concentration and temperature. For the classical regime to apply, this occupancy must be $\ll 1$. We note that ε_n as defined by Eq. (3.59) is always positive for a free atom.

Example: *N* atoms in a box

There follows now a tricky argument that we will use temporarily until we develop in Chapter 6 a powerful method to deal with the problem of many noninteracting identical atoms in a box. We first treat an ideal gas of N atoms in a box, all atoms of different species or different isotopes. This is a simple extension of the one atom result. We then discuss the major correction factor that arises when all atoms are identical, of the same isotope of the same species.

If we have one atomin each of N distinct boxes (see Figure 3.7), the partition function is the product of the separate one atom partition functions:

$$Z_{N\text{boxes}\%} = Z_1(1) Z_1(2) \cdots Z_1(N) \tag{3.66}$$

because the product on the right-hand side includes every independent state of the N boxes, such as the state of energy:

$$\varepsilon_\alpha(1) + \varepsilon_\beta(2) + \cdots + \varepsilon_\tau(N) \tag{3.67}$$

where $\alpha, \beta, \cdots, \tau$, denote the orbital indices of atoms in the successive boxes. The result of Eq. (3.66) also gives the partition function of N noninteracting atoms all of different species in a single box (see Figure 3.8):

$$Z_\alpha(\bullet) Z_\beta(\bigcirc) Z_\gamma(\blacktriangledown) \cdots Z_\tau(\triangle)$$

this being the same problem because the energy eigenvalues are the same as for Eq. (3.67).

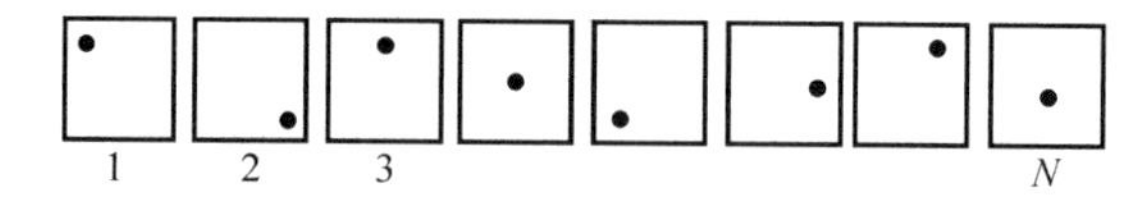

Figure 3.7 An N particle system of free particles with one particle in each of N boxes. The energy is N times that for one particle in one box

If the masses of all these different atoms happened to be the same, the total partition function would be Z_1^N, where Z_1 is given by Eq. (3.62).

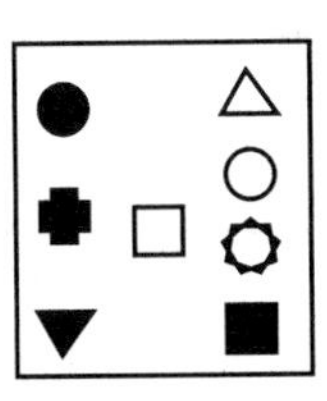

Figure 3.8 Atoms of different species in a single box

When we consider the more common problem of N identical particles in one box, we-have to correct $Z_1{}^N$ because it overcounts the distinct states of the N identical particle system. Particles of a single species are not distinguishable: electrons do not carry registration numbers. For two labeled particles ● and ▼ in a single box, the state ε_α(●)$+\varepsilon_\beta$(▼) and the state ε_α(▼)$+\varepsilon_\beta$(●) are distinct states, and both combinations must be counted in the partition function. But for two identical particles the state of energy $\varepsilon_\alpha+\varepsilon_\beta$ is the identical state as $\varepsilon_\beta+\varepsilon_\alpha$, and only one entry is to be made in the state sum in the partition function.

If the orbital indices are all different, each entry will oocur $N!$ times in $Z_1{}^N$, whereas the entry should occur only once if the particles are identical. Thus, $Z_1{}^N$ overcounts the states by a factor of $N!$, and the correct partition function for N identical partides is

$$Z_N = \frac{1}{N!}Z_1{}^N = \frac{1}{N!}(n_Q V)^N \tag{3.68}$$

in the classical regime. Here $n_Q=(M\tau/2\pi\hbar^2)^{3/2}$ from Eq. (3.63).

There is a step in the argument where we assume that all N occupied orbitals are always different orbitals. It is no simple matter to evaluate directly the error introduced by this approximation, but later we will confirm by another method the validity of Eq. (3.68) in the classical regime $n \ll n_Q$. The $N!$ factor changes the result for the entropy of the ideal gas. The entropy is an experimentally measurable quantity, and it has been confirmed that the $N!$ factor is correct in this low concentration limit.

Energy. The energy of the ideal gas follows from the N particle partition function by use of Eq. (3.12):

$$U = \tau^2(\partial \ln Z_N/\partial\tau) = \frac{3}{2}N\tau \tag{3.69}$$

consistent with Eq. (3.65) for one particle. The free energy is

$$F = -\tau\ln Z_N = -\tau\ln Z_1{}^N + \tau\ln N! \tag{3.70}$$

With the earlier result $Z_1 = n_Q V = (M\tau/2\pi\hbar^2)^{3/2}V$ and the Stirling approximation $\ln N! \approx N\ln N - N$, we have

$$F = -\tau N\ln\left[\left(\frac{M\tau}{2\pi\hbar^2}\right)^{\frac{3}{2}}V\right] + \tau N\ln N - \tau N \tag{3.71}$$

From the free energy we can calculate the entropy and the pressure of the ideal gas of N atoms. The pressure follows from Eq. (3.49):

$$p = -\left(\frac{\partial F}{\partial V}\right)_\tau = N\tau/V \tag{3.72}$$

or

$$pV = N\tau \tag{3.73}$$

which is called the ideal gas law. In conventional units,

$$pV = Nk_B T \tag{3.74}$$

The entropy follows from Eq. (3.49):

$$\sigma = -\left(\frac{\partial F}{\partial \tau}\right)_V = N\ln\left[\left(\frac{M\tau}{2\pi\hbar^2}\right)^{\frac{3}{2}} V\right] + \frac{3}{2}N - N\ln N + N \tag{3.75}$$

or

$$\sigma = N\left[\ln\left(\frac{n_Q}{n}\right) + \frac{5}{2}\right] \tag{3.76}$$

with the concentration $n \equiv N/V$. This result is known as the Sackur-Tetrode equation for the entropy of a monatomic ideal gas. It agrees with experiment. The result involves $\hbar$ through the term n_Q, so even for the classical ideal gas the entropy involves a quantum concept. We shall derive these results again in Chapter 6 by a direct method that does not explicitly involve the $N!$ or identical particle argument. The energy of Eq. (3.69) also follows from $U = F + \tau\sigma$; with use of Eq. (3.71) and Eq. (3.76) we have $U = \frac{3}{2}N\tau$.

Example: Equipartition of energy

The energy $U = \frac{3}{2}N\tau$ from Eq. (3.69) is ascribed to a contribution $\frac{1}{2}\tau$ from each "degree of freedom" of each particle, where the number of degrees of freedom is the number of dimensions of the space in which the atoms move: 3 in this example. In the classical form of statistical mechanics, the partition function contains the kinetic energy of the particles in an integral over the momentum components P_x, P_y, P_z. For one free particle

$$Z_1 = \iiint \exp[-(P_x^2 + P_y^2 + P_z^2)/2M\tau]\,dP_x\,dP_y\,dP_z \tag{3.77}$$

a result similar to Eq. (3.61). The limits of integration are $\pm\infty$ for each component. The thermal average energy may be calculated by use of Eq. (3.12) and is equal to $\frac{3}{2}\tau$.

The result is generalized in the classical theory. Whenever the hamiltonian of the system is homogeneous of degree 2 in a canonical momentum component, the classical limit of the thermal average kinetic energy associated with that momentum will be $\frac{1}{2}\tau$. Further, if the hamiltonian is homogeneous of degree 2 in a position coordinate component, the thermal average potential energy associated with that coordinate will also be $\frac{1}{2}\tau$. The result thus applies to the harmonic oscillator in the classical limit. At high temperatures the classical limits are attained, as in Figure 3.9.

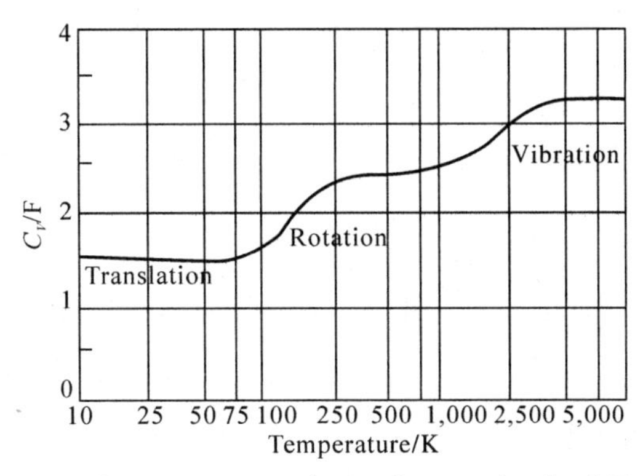

Figure 3.9 Heat capacity at constant volume of one molecule of H_2 in the gas phase

Example: Entropy of mixing

In Chapter 1 we calculated the number of possible arrangements of A and B in a solid made up of $N-t$ atoms A and t atoms B. We found in Eq. (1.20) for the number of arrangements:

$$g(N,t) = \frac{N!}{(N-t)!t!} \tag{3.78}$$

The entropy associated with these arrangements is

$$\sigma(N,t) = \ln g(N,t) = \ln N! - \ln(N-t)! - \ln t! \tag{3.79}$$

and is plotted in Figure 3.10 for $N=20$. This contribution to the total entropy of an alloy system is called the entropy of mixing. The result of Eq. (3.79) may be put in a more convenient form by use of the Stirling approximation:

$$\begin{aligned}\sigma(N,t) \approx\ & N\ln N - N - (N-t)\ln(N-t) + N - t - t\ln t + t = \\ & N\ln N - (N-t)\ln(N-t) - t\ln t = \\ & -(N-t)\ln\left(1-\frac{t}{N}\right) - t\ln(t/N)\end{aligned}$$

or, with $x \equiv t/N$,

$$\sigma(x) = -N[(1-x)\ln(1-x) + x\ln x] \tag{3.80}$$

This result gives the entropy of mixing of an alloy $A_{t-x}B_x$ treated as a random (homogeneous) solid solution.

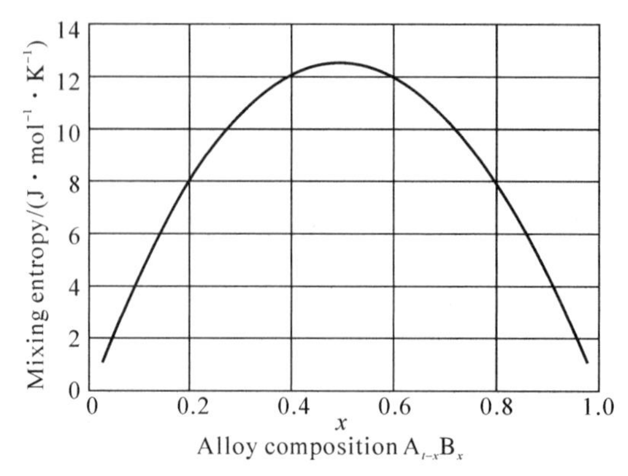

Figure 3. 10 Mixing entropy of a random binary alloy as a function of the proportions of the constituent atoms A and B

We ask: Is the homogeneous solid solution the equilibrium condition of a mixture of A and B atoms, or is the equilibrium a two-phase system, such as a mixture of crystallites of pure A and crystallites of pure B? The complete answer is the basis of much of the science of metallurgy: the answer will depend on the temperature and on the interatomic inter—action energies U_{AA}, U_{BB}, and U_{AB}. In the special case that the interaction energies between AA, BB, and AB neighbor pairs are all equal, the homogeneous solid solution will have a lower free energy than the corresponding mixture of crystallites of the pure elements. The free energy of the solid solution $A_{l-x}B_x$ is

$$F = F_0 - \tau\sigma(x) = F_0 + N\tau[(1-x)\ln(1-x) + x\ln x] \tag{3.81}$$

which we must compare with

$$F = (1-x)F_0 + xF_0 = F_0 \tag{3.82}$$

for the mixture of A and B crystals in the proportion $(1-x)$ to x. The entropy of mixing is always positive—all entropies are positive—so that the solid solution has the lower free energy in this special case.

There is a tendency for at least a very small proportion of any element B to dissolve in any other element A, even if a strong repulsive energy exists between a B atom and the surrounding A atoms. Let this repulsive energy be denoted by U, a positive quantity. If a very small proportion $x \ll 1$ of B atoms is present, the total repulsive energy is xNU, where xN is the number of B atoms. The mixing entropy of Eq. (3.80) is approximately

$$\sigma = -xN\ln x \tag{3.83}$$

in this limit, so that the free energy is

$$F(x) = N(xU + \tau x\ln x) \tag{3.84}$$

which has a minimum when

$$\frac{\partial F}{\partial x} = N(U + \tau \ln x + \tau) = 0 \tag{3.85}$$

or

$$x = \exp(-1)\exp(-U/\tau) \tag{3.86}$$

This shows there is a natural impurity content in all crystals.

3.8 Summary

(1) The factor

$$P(\varepsilon_s) = \exp\left(-\frac{\varepsilon_s}{\tau}\right)/Z$$

is the probability of finding a system in a state s of energy ε_s when the system S is a positive integer or zero, and ω is the classical frequency of the oscillator. We have chosen the zero of energy at the state $s=0$.

1) Show that for a harmonic oscillator the free energy is

$$F = \tau \ln[1 - \exp(-\hbar\omega/\tau)] \tag{3.87}$$

Note that at high temperatures such that $\tau \gg \hbar\omega$ we may expand the argument of the logarithm to obtain $F \approx \tau \ln(\hbar\omega/\tau)$.

2) From Eq. (3.87) show that the entropy is

$$\sigma = \frac{\hbar\omega/\tau}{\exp\left(\frac{\hbar\omega}{\tau}\right) - 1} - \ln[1 - \exp(-\hbar\omega/\tau)] \tag{3.88}$$

The entropy is shown in Figure 3.11 and the heat capacity in Figure 3.12.

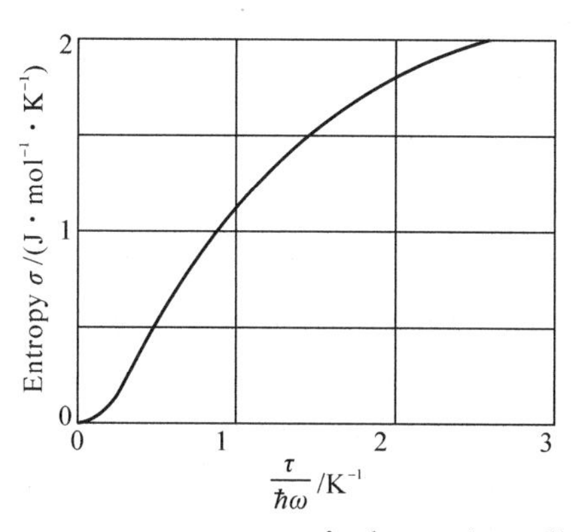

Figure 3.11 Entropy versus temperature for harmonic oscillator of frequency ω

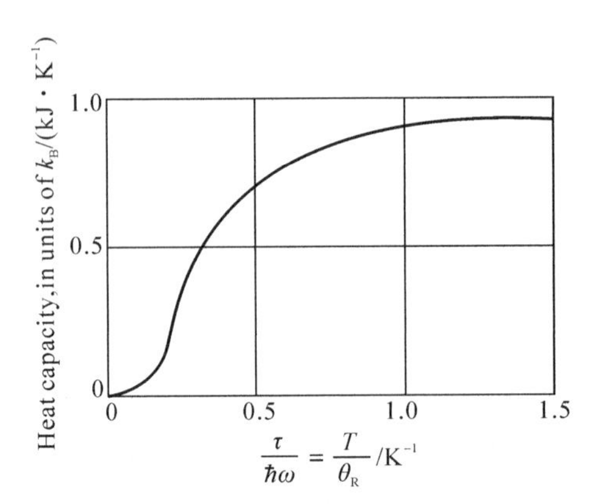

Figure 3.12 Heat capacity versus temperature for harmonic oscillator of frequency ω

(2) Energy fluctuations. Consider a system of fixed volume in thermal contact with a reservoir. Show that the mean square fluctuation in the energy of the system is

$$< (\varepsilon - < \varepsilon >)^2 > = \tau^2 (\partial U / \partial \tau)_V \tag{3.89}$$

Here U is the conventional symbol for $<\varepsilon>$. Hint: Use the partition function Z to relate $\partial U/\partial \tau$ to the mean square fluctuation. Also, multiply out the term $(\cdots)^2$. Note: The temperature τ of a system is a quantity that by definition does not fluctuate in value when the system is in thermal contact with a reservoir. Any other attitude would be inconsistent with our definition of the temperature of a system. The energy of such a system may fluctuate, but the temperature does not. Some workers do not adhere to a rigorous definition of temperature. Thus Landau and Lifshitz give the result

$$< (\Delta\tau)^2 > = \tau^2 / C_V \tag{3.90}$$

but this should be viewed as just another form of (3.89) with $\Delta\tau$ set equal to $\Delta U/C_V$. We know that $\Delta U = C_V \Delta\tau$, whence Eq. (3.90) becomes $<(\Delta U)^2> = \tau^2 C_V$, which is our result of Eq. (3.89).

(3) Overhauser effect. Suppose that by a suitable external mechanical or electrical arrangement one can add $\alpha\varepsilon$ to the energy of the heat reservoir whenever the reservoir passes to the system the quantum of energy ε. The net increase of energy of the reservoir is $(\alpha - 1)\varepsilon$. Here α is some numerical factor, positive or negative. Show that the effective Boltzmann factor for this abnormal system is given by

$$P(\varepsilon) = \exp[-(1-\alpha)\varepsilon/\tau] \tag{3.91}$$

This reasoning gives the statistical basis of the Overhauser effect whereby the nuclear polarization in a magnetic field can be enhanced above the thermal equilibrium polarization. Such a condition requires the active supply of energy to the system from an extemal source. The system is not in equilibrium, but is said to be in a steady state. Cf. A. W. Overhauser,

Phys. Rev. 92. 411 (1953).

(4) Rotation of diatomic molecales. In our first look at the ideal gas we considered only the translational energy of the particles. But molecules can rotate, with kinetic energy. The rotational motion is quantized; and the energy levels of a diatomic molecule are of the form

$$\varepsilon(j) = j(j+1)\varepsilon_0 \tag{3.92}$$

where j is any Positive integer including zero: $j=0,1,2,\cdots$. The multiplicity of each rotational level is $g(j)=2j+1$.

1) Find the partition function $Z_R(\tau)$ for the rotational states of one molecule. Remember that Z is a sum over all states, not over all levels—this makes a difference.

2) Evaluate $Z_R(\tau)$ approximately for $\tau \gg \varepsilon_0$, by converting the sum to an integral.

3) Do the same for $\tau \gg \varepsilon_0$, by truncating the sum after the second term.

4) Give expressions for the energy U and the heat capacity C, as functions of τ, in both limits. Observe that the rotational contribution to the heat capacity of a diatomic molecule approaches 1 (or, in conventional units, k_B) when $\tau \gg \varepsilon_0$.

5) Sketch the behavior of $U(\tau)$ and $C(\tau)$, showing the limiting behaviors for $\tau \to \infty$ and $\tau \to 0$.

(5) Zipper problem. A zipper has N links; each link has a state in which it is closed with energy 0 and a state in which it is open with energy ε. We require, however, that the zipper can only unzip from the left end, and that the link number s can only open if all links to the left $(1,2,\cdots,s-1)$ are already open.

1) Show that the partition function can be summed in the form

$$Z = \frac{1-\exp[-(N+1)\varepsilon/\tau]}{1-\exp(-\varepsilon/\tau)} \tag{3.93}$$

2) In the limit $\varepsilon \gg \tau$, find the average number of open links. The model is a very simplified model of the unwinding of two-stranded DNA molecules.

(6) Quantum concentration. Consider one particle confined to a cube of side L; the concentration in effect is $n=1/L^3$. Find the kinetic energy of the particle when in the ground orbital. There will be a value of the concentration for which this zero—point quantum kinetic energy is equal to the temperature τ(At this concentration the occupancy of the lowest orbital is of the order of unity; the lowest orbital always has a higher occupancy than any other orbital.). Show that the concentration n_0 thus defined is equal to the quantum concentration n_Q defined by Eq. (3.63), within a factor of the order of unity.

(7) Partition function for two systems. Show that the partition function $Z(1+2)$ of two independent systems 1 and 2 in thermal contact at a common temperature τ is equal to the product of the partition functions of the separate systems:

$$Z(1+2) = Z(1)Z(2) \tag{3.94}$$

(8) Elasticity of polymers. The thermodynamic identity for a one-dimensional system is

$$\tau d\sigma = dU - f dl \tag{3.95}$$

when f is the external force exerted on the line and dl is the extension of the line. By analogy with Eq. (3.32) we form the derivative to find

$$-\frac{f}{\tau} = \left(\frac{\partial \sigma}{\partial l}\right)_U \tag{3.96}$$

The direction of the force is opposite to the conventional direction of the pressure.

We consider a polymeric chain of N links each of length p, with each link equally likely to be directed to the right and to the left.

1) Show that the number of arrangements that give a head-to-tail length of $l=2|s|\rho$ is

$$g(N,-s)+g(N,s) = \frac{2N!}{\left(\frac{1}{2}N+s\right)!\left(\frac{1}{2}N-s\right)!} \tag{3.97}$$

2) For $|s| \ll N$ show that

$$\sigma(l) = \ln\left[2g(N,0)\right] - l^2/2N\rho^2 \tag{3.98}$$

3) Show that the force at extension l is

$$f = l\tau/N\rho^2 \tag{3.99}$$

The force is proportional to the temperature. The force arises because thepolymer wants to curl up: the entropy is higher in a random coil than in an uncoiled configuration. Warming a rubber band makes it contract; warming a steel wire makes it expand. The theory of rubber elasticity is discussed by H. M. James and E. Guth, Journal of Chemical Physics 11, 455 (1943); Journal of Polymer Science 4, 153 (1949); see also L. R. G. Treloar, Physics of rubber elasticity, Oxford, 1958.

(9) One-dimensional gas. Consider an ideal gas of N particles, each of mass M, confined to a one-dimensional line of length L. Find the entropy at temperature τ. The particles have spin zero.

Chapter 4 Thermal Radiation and Planck Distribution

4.1 Planck Distribution Function

The Planck distribution describes the spectrum of the electromagnetic radiation in thermal equilibrium within a cavity. Approximately, it describes the emission spectrum of the Sun or of metal heated by a welding torch. The Planck distribution was the first application of quantum thermal physics. Thermal electromagnetic radiation is often called black body radiation. The Planck distribution also describes the thermal energy spectrum of lattice vibrations in an elastic solid.

The word "mode" characterizes a particular oscillation amplitude pattern in the cavity or in the solid. We shall always refer to $\omega=2\pi f$ as the frequency of the radiation. The characteristic feature of the radiation problem is that a mode of oscillation of frequency ω may be excited only in units of the quantum of energy $\hbar\omega$. The energy ε_s of the state with s quanta in the mode is

$$\varepsilon_s = s\hbar\omega \tag{4.1}$$

where s is zero or any positive integer (see Figure 4.1). We omit the zero point energy $\frac{1}{2}\hbar\omega$.

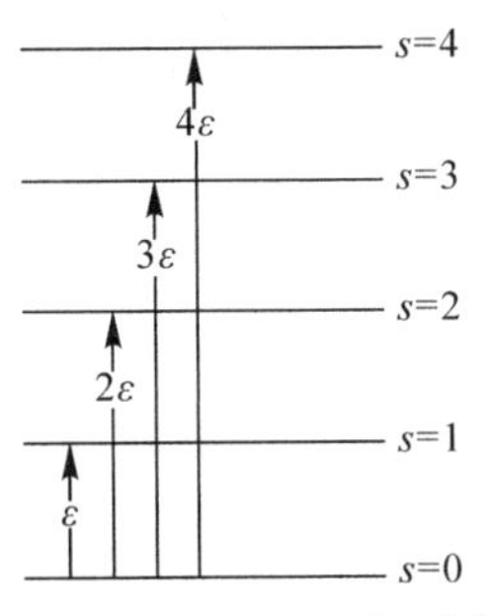

Figure 4.1 States of an oscillator that represents a mode of frequency ω of an electromagnetic field

These energies are the same as the energies of a quantum harmonic oscillator of frequency ω, but there is a difference between the concepts. A harmonic oscillator is a localized oscillator, whereas theelectric and magnetic energy of an electromagnetic cavity mode is distributed throughout the interior of the cavity (see Figure 4. 2). For both problems the energy eigenvalues are integral multiples of $\hbar\omega$, and this is the reason for the similarity in the thermal physics of the two problems. The language used to describe an excitation is different: s for the oscillator is called the quantum number, and s for the quantized electromagnetic mode is called the number of photons in the mode.

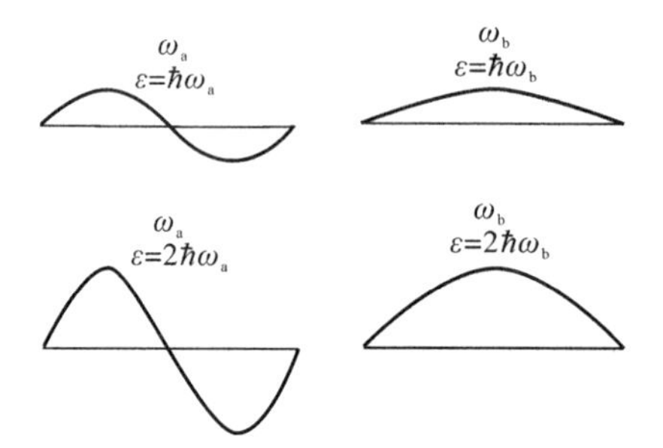

Figure 4. 2 Representation in one dimension of two electromagnetic modes a and b, of frequency ω_a and ω_b

We first calculate the thermal average of the number of photons in a mode, when these photons are in thermal equilibrium with a reservoir at a temperature τ. The partition function of Eq. (3. 10) is the sum over the states of Eq. (4. 1):

$$Z=\sum_{s=0}^{\infty}\exp(-s\hbar\omega/\tau) \tag{4. 2}$$

This sum is of the form $\sum x^s$, with $x\equiv\exp\left(-\frac{\hbar\omega}{\tau}\right)$. Because x is smaller than 1, the infinite series may be summed and has the value $1/(1-x)$, whence

$$Z=\frac{1}{1-\exp\left(-\frac{\hbar\omega}{\tau}\right)} \tag{4. 3}$$

The probability that the system is in the state s of energy $s\hbar\omega$ is given by the Boltzmann factor:

$$P(s)=\frac{\exp\left(-\frac{s\hbar\omega}{\tau}\right)}{Z} \tag{4. 4}$$

The thermal average value of s is

$$<s>=\sum_{s=0}^{x}sP(s)=Z^{-1}\sum_{s=0}^{x}s\exp\left(-\frac{s\hbar\omega}{\tau}\right) \tag{4. 5}$$

With $y\equiv\frac{\hbar\omega}{\tau}$, the summation on the right-hand side has the form:

$$\sum s\exp(-sy)=-\frac{\mathrm{d}}{\mathrm{d}y}\sum\exp(-sy)=-\frac{\mathrm{d}}{\mathrm{d}y}\left[\frac{1}{1-\exp(-y)}\right]=\frac{\exp(-y)}{[1-\exp(-y)]^2}$$

from Eq. (4. 3) and Eq. (4. 5) we find

$$< s >= \frac{\exp(-y)}{1-\exp(-y)}$$

or

$$< s >= \frac{1}{\exp\left(\frac{\hbar\omega}{\tau}\right)-1} \tag{4.6}$$

This is the Planck distribution function for the thermal average number of photons (see Figure 4.3) in a single mode of frequency ω. Equally, it is the average number of phonons in the mode. The result applies to any kind of wave field with energy in the form of Eq. (4.1).

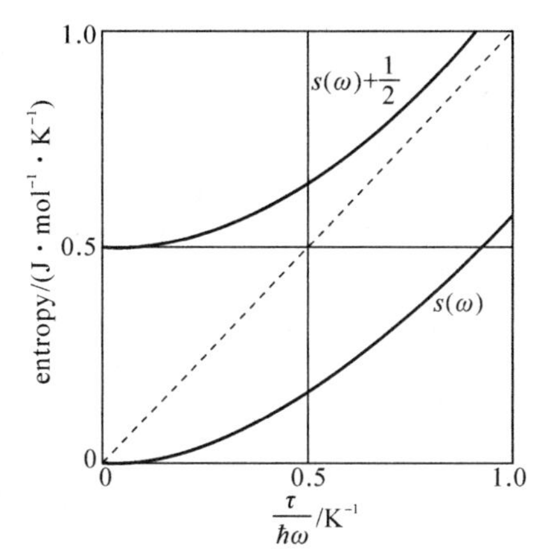

Figure 4.3 Planck distribution as a function of the reduced temperature $\tau/\hbar\omega$

4.2 Planck Law and Stefan-Boltzmann Law

The thermal average energy in the mode is

$$< \varepsilon > = < s > \hbar\omega = \frac{\hbar\omega}{\exp\left(\frac{\hbar\omega}{\tau}\right)-1} \tag{4.7}$$

Here $<s(\omega)>$ is the thermal average of the number of photons in the mode of frequency ω. A plot of $<s(\omega)>+\frac{1}{2}$ is also given, where $\frac{1}{2}$ is the effective zero point occupancy of the mode; the dashed line is the classical asymptote. Note that we write

$$< s >+\frac{1}{2} = \frac{1}{2}\coth\left(\frac{\hbar\omega}{2\tau}\right)$$

The high temperature limit $\tau \gg \hbar\omega$ is often called the classical limit. Here $\exp\left(\frac{\hbar\omega}{\tau}\right)$ may be approximated as $1+\frac{\hbar\omega}{\tau}+\cdots$, whence the classical average energy is

$$< \varepsilon > \approx \tau \tag{4.8}$$

There is an infinite number of electromagnetic modes within any cavity. Each mode $\boldsymbol{n}$ has its own frequency ω_n. For radiation confined within a perfectly conducting cavity in the form of a cube of edge L, there is a set of modes of the form

$$\boldsymbol{E}_x = \boldsymbol{E}_{x0} \sin\omega t \cos\left(\frac{\boldsymbol{n}_x \pi x}{L}\right) \sin\left(\frac{\boldsymbol{n}_y \pi y}{L}\right) \sin\left(\frac{\boldsymbol{n}_z \pi z}{L}\right) \tag{4.9a}$$

$$\boldsymbol{E}_y = \boldsymbol{E}_{y0} \sin\omega t \cos\left(\frac{\boldsymbol{n}_x \pi x}{L}\right) \sin\left(\frac{\boldsymbol{n}_y \pi y}{L}\right) \sin\left(\frac{\boldsymbol{n}_z \pi z}{L}\right) \tag{4.9b}$$

$$\boldsymbol{E}_z = \boldsymbol{E}_{z0} \sin\omega t \cos\left(\frac{\boldsymbol{n}_x \pi x}{L}\right) \sin\left(\frac{\boldsymbol{n}_y \pi y}{L}\right) \sin\left(\frac{\boldsymbol{n}_z \pi z}{L}\right) \tag{4.9c}$$

Here $\boldsymbol{E}_x$, $\boldsymbol{E}_y$ and $\boldsymbol{E}_z$ are the three electric field components, and $\boldsymbol{E}_{x0}$, $\boldsymbol{E}_{y0}$ and $\boldsymbol{E}_{z0}$ are the corresponding amplitudes. The three components are not independent, because the field must be divergence-free:

$$\mathrm{div}\boldsymbol{E} = \frac{\partial \boldsymbol{E}_x}{\partial x} + \frac{\partial \boldsymbol{E}_y}{\partial y} + \frac{\partial \boldsymbol{E}_z}{\partial z} = 0 \tag{4.10}$$

When we insert Eq. (4.9) into Eq. (4.10) and drop all common factors, we find the condition

$$\boldsymbol{E}_{x0}\boldsymbol{n}_x + \boldsymbol{E}_{y0}\boldsymbol{n}_y + \boldsymbol{E}_{z0}\boldsymbol{n}_z = \boldsymbol{E}_0\boldsymbol{n} = 0 \tag{4.11}$$

This states that the field vectors must be perpendicular to the vector $\boldsymbol{n}$ with the components $\boldsymbol{n}_x$, $\boldsymbol{n}_y$ and $\boldsymbol{n}_z$, so that the electromagnetic field in the cavity is a transversely polarized field. The polarization direction is defined as the direction of $\boldsymbol{E}_0$.

For a given triplet $\boldsymbol{n}_x$, $\boldsymbol{n}_y$, $\boldsymbol{n}_z$, we can choose two mutually perpendicular polarization directions, so that there are two distinct modes for each triplet $\boldsymbol{n}_x$, $\boldsymbol{n}_y$, $\boldsymbol{n}_z$.

On substitution of Eq. (4.9) in the wave equation

$$c^2\left(\frac{\partial^2}{\partial x^2} + \frac{\partial^2}{\partial y^2} + \frac{\partial^2}{\partial z^2}\right)\boldsymbol{E}_z = \frac{\partial^2 \boldsymbol{E}_z}{\partial t^2} \tag{4.12}$$

with c the velocity of light, we find

$$c^2\pi^2(\boldsymbol{n}_x{}^2 + \boldsymbol{n}_y{}^2 + \boldsymbol{n}_z{}^2) = \omega^2 L^2 \tag{4.13}$$

This determines the frequency ω of the mode in terms of the triplet of integers $\boldsymbol{n}_x$, $\boldsymbol{n}_y$, $\boldsymbol{n}_z$. If we define

$$\boldsymbol{n} \equiv (\boldsymbol{n}_x{}^2 + \boldsymbol{n}_y{}^2 + \boldsymbol{n}_z{}^2)^{1/2} \tag{4.14}$$

then the frequencies are of the form

$$\omega_n = \boldsymbol{n}\pi c/L \tag{4.15}$$

The total energy of the photons in the cavity is, from Eq. (4.7),

$$U = \sum_n < \varepsilon_n > = \sum_n \frac{\hbar\omega_n}{\exp\left(\frac{\hbar\omega_n}{\tau}\right) - 1} \tag{4.16}$$

The sum is over the triplet of integers $\boldsymbol{n}_x$, $\boldsymbol{n}_y$, $\boldsymbol{n}_z$. Positive integers alone will describe all independent modes of the form Eq. (4.9). We replace the sum over $\boldsymbol{n}_x$, $\boldsymbol{n}_y$, $\boldsymbol{n}_z$ by an integral over the volume element $\mathrm{d}\boldsymbol{n}_x \mathrm{d}\boldsymbol{n}_y \mathrm{d}\boldsymbol{n}_z$ in the space of the mode indices.

That is, we set

$$\sum_n (\cdots) = \frac{1}{8}\int_0^x 4\pi \boldsymbol{n}^2 \mathrm{d}\boldsymbol{n}(\cdots) \tag{4.17}$$

where the factor $\frac{1}{8}=\left(\frac{1}{2}\right)^3$ arises because only the positive octant of the space is involved. We now multiply the sum or integral by a factor of 2 because there are two independent polarizations of the electromagnetic field (two independent sets of cavity modes). Thus

$$U = \pi\int_0^\infty \mathrm{d}\boldsymbol{n}\boldsymbol{n}^2 \frac{\hbar\omega_n}{\exp\left(\frac{\hbar\omega_n}{\tau}\right)-1} = (\pi^2\hbar c/L)\int_0^\infty \mathrm{d}\boldsymbol{n}\boldsymbol{n}^3 \frac{1}{\exp\left(\frac{\hbar c\boldsymbol{n}\pi}{L\tau}\right)-1} \tag{4.18}$$

with Eq. (4.15) for ω_n. Standard practice is to transform the definite integral to one over a dimensionless variable. We set $x\equiv\frac{\hbar c\boldsymbol{n}\pi}{L\tau}$, and Eq. (4.18) becomes

$$U = (\pi^2\hbar c/L)(L\tau/\hbar c\pi)^4\int_0^\infty \mathrm{d}x \frac{x^3}{\exp x - 1} \tag{4.19}$$

The definite integral has the value $\pi^4/15$; it is found in good standard tables such as Dwight (cited in the general references). The energy per unit volume is

$$\frac{U}{V} = \frac{\pi^2}{15\hbar^3 c^3}\tau^4 \tag{4.20}$$

with the volume $V=L^3$. The result that the radiant energy density is proportional to the fourth power of the temperature is known as the Stefan-Boltzmann law of radiation.

For many applications of this theory we decompose Eq. (4.20) into the spectral density of the radiation. The spectral density is defined as the energy per unit volume per unit frequency range, and is denoted as u_ω. We can find u_ω from Eq. (4.18) rewritten in terms of ω:

$$U/V = \int \mathrm{d}\omega u_\omega = \frac{\hbar}{\pi^2 c^3}\int \mathrm{d}\omega \frac{\omega^3}{\exp(\hbar\omega/\tau)-1} \tag{4.21}$$

so that the spectral density is

$$u_\omega = \frac{\hbar}{\pi^2 c^3}\frac{\omega^3}{\exp(\hbar\omega/\tau)-1} \tag{4.22}$$

This function is involved in the Planck radiation law for the spectral density u_ω. The temperature of a black body may be found from the frequency $\omega_{\max}$ at which the radiant energy density is a maximum, per unit frequency range. This frequency is directly proportional to the temperature.

This result is the Planck radiation law; it gives the frequency distribution of thermal radiation (see Figure 4.4). Quantum theory began here.

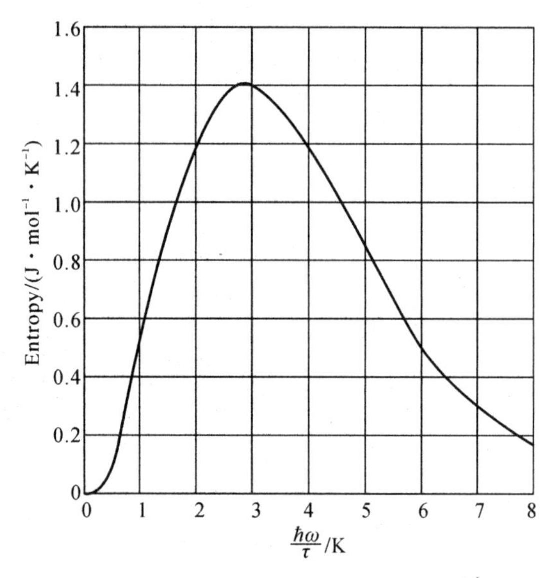

Figure 4.4 Plot of $x^3/(e^x-1)$ with $x \equiv \frac{\hbar\omega}{\tau}$

The entropy of the thermal photons can be found from the relation of Eq. (3.34a) at constant volume: $d\sigma = dU/\tau$, whence from Eq. (4.20),

$$d\sigma = \frac{4\pi^2 V}{15\hbar^3 c^3}\tau^2 d\tau$$

Thus the entropy is

$$\sigma(\tau) = (4\pi^2 V/45)(\tau/\hbar c)^3 \tag{4.23}$$

The constant of integration is zero, from Eq. (3.55) and the relation between F and σ.

A process carried out at constant photon entropy will have $V\tau^3$ = constant.

The measurement of high temperatures depends on the flux of radiant energy from a small hole in the wall of a cavity maintained at the temperature of interest. Such a hole is said to radiate as a black body—which means that the radiation emission is characteristic of a thermal equilibrium distribution. The energy flux density J_U is defined as the rate of energy emission per unit area. The flux density is of the order of the energy contained in a column of unit area and length equal to the velocity of light times the unit of time. Thus,

$$J_U = [cU(\tau)/V] \times (\text{geometrical factor}) \tag{4.24}$$

The geometrical factor is equal to 1/4; the derivation is the subject of Problem 15.

The final result for the radiant energy flux is

$$J_U = \frac{cU(\tau)}{4V} = \frac{\pi^2 \tau^4}{60\hbar^3 c^2} \tag{4.25}$$

by use of Eq. (4.20) for the energy density $\frac{U}{V}$. The result is often written as

$$J_U = \sigma_B T^4 \tag{4.26a}$$

1. The Stefan-Boltzmann constant

$$\sigma_B \equiv \pi^2 k_B{}^4 / 60\hbar^3 c^2 \tag{4.26b}$$

has the value 5.670×10^{-8} W · m^{-2} · K^{-4} or 5.670×10^{-5} erg · cm^{-2} · s^{-1} · K^{-4} (Here σ_B is not the entropy.). A body that radiates at this rate is said to radiate as a black body. A small hole in a cavity whose walls are in thermal equilibrium at temperature T will radiate as a black body at the rate given in Eq. (4.26a). The rate is independent of the physical constitution of the walls of the cavity and depends only on the temperature.

2. Emission and absorption: Kirchhoff law

The ability of a surface to emit radiation is proportional to the ability of the surface to absorb radiation. We demonstrate this relation, first for a black body or black surface and, second, for a surface with arbitrary properties. An object is defined to be black in a given frequency range if all electromagnetic radiation incident upon it in that range is absorbed. By this definition a hole in a cavity is black if the hole is small enough that radiation incident through the hole will reflect enough times from the cavity walls to be absorbed in the cavity with negligible loss back through the hole.

The radiant energy flux density J_U from a black surface at temperature τ is equal to the radiant energy flux density J_U emitted from a small hole in a cavity at the same temperature. To prove this, let us close the hole with the black surface, hereafter called the object. In thermal equilibrium the thermal average energy flux from the black object to the interior of the cavity must be equal, but opposite, to the thermal average energy flux from the cavity to the black object.

We prove the following: If a non-black object at temperature τ absorbs a fraction a of the radiation incident upon it, the radiation flux emitted by the object will be a times the radiation flux emitted by a black body at the same temperature. Let a denote the absorptivity and e the emissivity, where the emissivity is defined so that the radiation flux emitted by the object is e times the flux emitted by a black body at the same temperature. The object must emit at the same rate as it absorbs if equilibrium is to be maintained. It follows that $a = e$. This is the Kirchhoff law. For the special case of a perfect reflector, a is zero, whence e is zero. A perfect reflector does not radiate.

The arguments can be generalized to apply to the radiation at any frequency, as between

ω and $\omega + d\omega$. We insert a filter between the object and the hole in the black body. Let the filter reflect perfectly outside this frequency range, and let it transmit perfectly within this range. The flux equality arguments now apply to the transmitted spectral band, so that $a(\omega) = e(\omega)$ for any surface in thermal equilibrium.

3. Estimation of surface temperature

One way to estimate the surface temperature of a hot body such as a star is from the frequency at which the maximum emission of radiant energy takes place (see Figure 4.4). What this frequency is depends on whether we look at the energy flux per unit frequency range or per unit wavelength range. For it u_ω, the energy density per unit frequency range, the maximum is given from the Planck law, Eq. (4.22), as

$$\frac{d}{dx}\left(\frac{x^3}{\exp x - 1}\right) = 0$$

or

$$3 - 3\exp(-x) = x$$

This equation may be solved numerically. The root is

$$\hbar\omega_{\max}/k_B T = x_{\max} \approx 2.82 \tag{4.27}$$

as in Figure 4.4.

Example: Cosmic black body background radiation

A major recent discovery is that the universe accessible to us is filled with radiation approximately like that of a black body at 2.9 K. The existence of this radiation (see Figure 4.5) is important evidence for big bang cosmological models which assume that the universe is expanding and cooling with time. This radiation is left over from an early epoch when the universe was composed primarily of electrons and protons at a temperature of about 4,000 K. The plasma of electrons and protons interacted strongly with electromagnetic radiation at all important frequencies, so that the matter and the black body radiation were in thermal equilibrium. By the time the universe had cooled to 3,000 K, the matter was primarily in the form of atomic hydrogen. This interacts with black body radiation only at the frequencies of the hydrogen spectral lines. Most of the black body radiation energy thus was effectively decoupled from the matter. Thereafter the radiation evolved with time in a very simple way: the photon gas was cooled by expansion at constant entropy to a temperature of 2.9 K. The photon gas will remain at constant entropy if the frequency of each mode is lowered during the expansion of the universe with the number of photons in each mode kept constant. We show in Eq. (4.58) below that the entropy is constant if the number of photons in each mode is constant-the occupancies determine the entropy.

After the decoupling the evolution of matter into heavier atoms (which are organized into galaxies, stars, and dust clouds) was more complicated than before decoupling. Electromagnetic radiation, such as starlight, radiated by the matter since the decoupling is superimposed on the cosmic black body radiation.

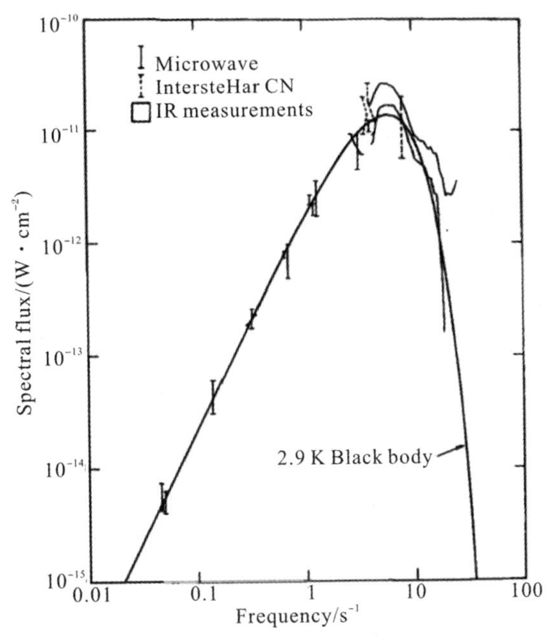

Figure 4. 5 Experimental measurements of the spectrum of the cosmic black body radiation

4. 3 Electrical Noise

As an important example of the Planck law in one dimension, we consider the spontaneous thermal fluctuations in voltage across a resistor. These fluctuations, which are called noise, were discovered by J. B. Johnson and explained by H. Nyquist. The characteristic property of Johnson noise is that the mean square noise voltage is proportional to the value of the resistance R, as shown by Figure 4. 6. We shall see that $\langle V^2 \rangle$ is also direct proportional to the temperature τ and the bandwidth Δf of the circuit(This section presumes a knowledge of electromagnetic wave propagation at the intermediate level.).

The Nyquist theorem gives a quantitative expression for the thermal noise voltage generated by a resistor in thermal equilibrium. The theorem is therefore needed in any estimate of the limiting signal-to-noise ratio of an experimental.

In the original form the Nyquist theorem states that the mean square voltage across a resistor of resistance R in thermal equilibrium at temperature τ is given by

$$\langle V^2 \rangle = 4R\tau\Delta f \tag{4.28}$$

where Δf is the frequency bandwidth within which the voltage fluctuations are measured; all frequency components outside the given range are ignored. We show below that the thermal noise power per unit frequency range delivered by a resistor to a matched load is τ; the factor 4 enters where it does because in the circuit of Figure 4.7, the power delivered to an arbitrary resistive load R' is

$$\langle I^2 R' \rangle = \frac{\langle V^2 R' \rangle}{(R+R')^2} \tag{4.29}$$

which at match ($R'=R$) is $\langle V^2 \rangle / 4R$.

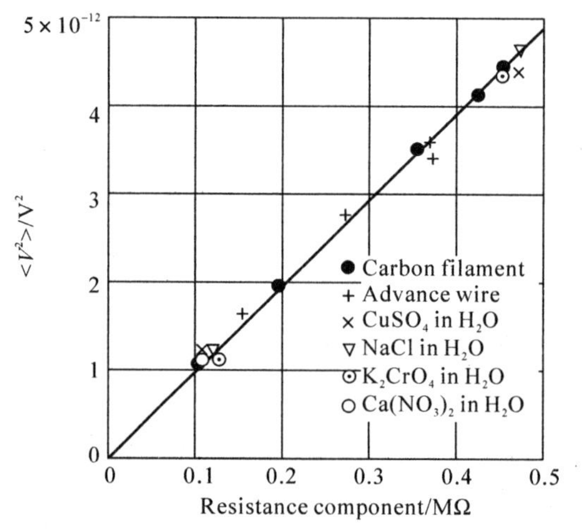

Figure 4.6 Voltage squared versus resistant for various kinds of conductors, including electrolytes

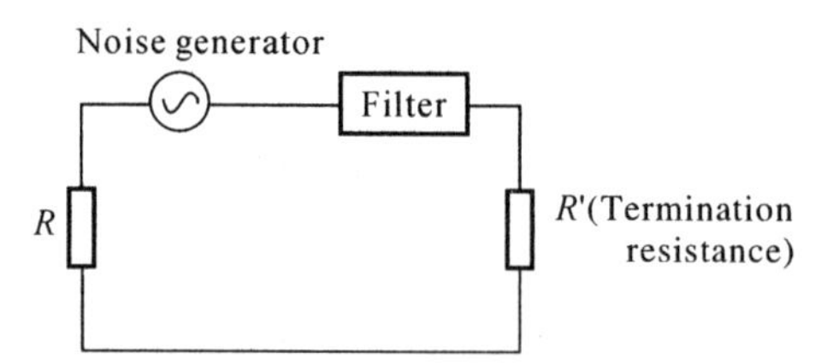

Figure 4.7 Equivalent circuit for a resistance R with a generator of thermal noise that delivers power to a load R'

The current

$$I = \frac{V}{R+R'}$$

so that the mean power dissipated in the load is

$$\varphi = \langle I^2 \rangle R' = \frac{\langle V^2 R' \rangle}{(R+R')^2}$$

which is a maximum with respect to R' when $R'=R$. In this condition the load is said to be matched to the power supply. At match, $\varphi=\langle V^2\rangle/4R$. The filter enables us to limit the frequency bandwidth under consideration; that is the bandwidth to which the mean square voltage fluctuation applies.

Consider as in Figure 4.8 a lossless transmission line of length L and characteristic impedance $Z_c=R$ terminated at each end by a resistance R. Thus the line is matched at each end, in the sense that all energy traveling down the line will be absorbed without reflection in the appropriate resistance. The entire circuit is maintained at temperature τ.

A transmission line is essentially an electromagnetic system in one dimension. We follow the argument given above for the distribution of photons in thermal equilibrium, but now in a space of one dimension instead of three dimensions. The transmission line has two photon modes (one propagating in each direction) of frequency $2\pi f_n=2n\pi/L$ from Eq. (4.15), so that there are two modes in the frequency range:

$$\delta f = c'/L \tag{4.30}$$

where c' is the propagation velocity on the line. Each mode has energy

$$\frac{\hbar\omega}{\exp\left(\frac{\hbar\omega}{\tau}\right)-1} \tag{4.31}$$

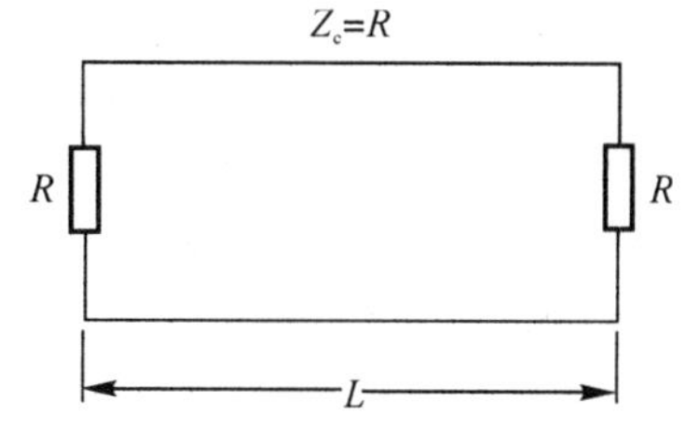

Figure 4.8　Transmission line of length L with matched terminations, as conceived for the derivation of the Nyquist theorem

In equilibrium, according to the Planck distribution. We are usually concerned with circuits in the classical limit $\hbar\omega\ll\tau$ so that the thermal energy per mode is τ. It follows that the energy on the line in the frequency range Δf is

$$\frac{2\tau\Delta f}{\delta f}=\frac{2\tau L\Delta f}{c'} \tag{4.32}$$

The rate at which energy comes off the line in one direction is $\tau\Delta f$.

The power coming off the line at one end is all absorbed in the terminal impedance R at that end; there are no reflections when the terminal impedance is matched to the line. In thermal equilibrium the load must emit energy to the line at the same rate, or else its temperature would rise. Thus the power input to the load is

$$\varphi = \langle I^2 \rangle R = \tau \Delta f \tag{4.33}$$

but $V=2RI$, so that Eq. (4.28) is obtained. The result has been used in low temperature thermometry in temperature regions (see Figure 4.9) where it is more convenient to measure $\langle V^2 \rangle$ than τ. Johnson noise is the noise across a resistor when no do current is flowing. Additional noise (not discussed here) appears when a do current flows.

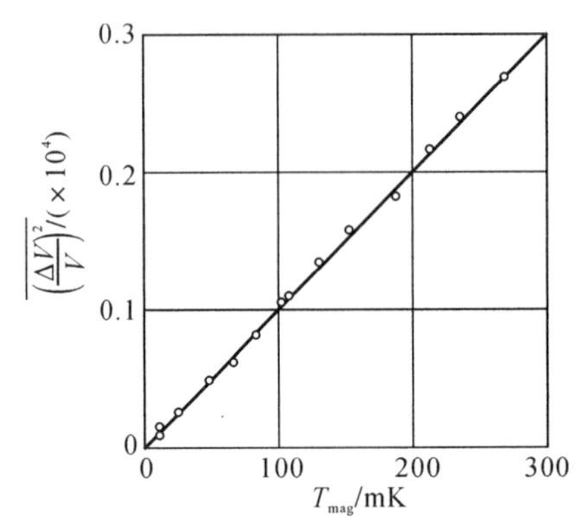

Figure 4.9 Mean square noise voltage fluctuations observed experiments

4.4 Phonons in Solids: Debye Theory

The energy of an elastic wave in a solid is quantized just as the energy of an electromagnetic wave in a cavity is quantized. The quantum of energy of an elastic wave is called a phonon. The thermal average number of phonons in an elastic wave of frequency ω is given by the Planck distribution function, just as for photons:

$$\langle s(\omega) \rangle = \frac{1}{\exp(\hbar\omega/\tau) - 1} \tag{4.34}$$

We assume that the frequency of an elastic waveis independent of the amplitude of the elastic strain. We want to find the energy and heat capacity of the elastic waves in solids. Several of the results obtained for photons may be carried over to phonons. The results are simple if we assume that the velocities of all elastic waves are equal—independent of frequency, direction of propagation, and direction of polarization. This assumption is not very accurate, but it helps account for the general trend of the observed results in many solids, with a minimum of computation.

There are two important features of the experimental results: the heat capacity of a nonmetallic solid varies as τ^3 at low temperatures, and at high temperatures the heat capacity is independent of the temperature.

1. Number of phonon modes

There is no limit to the number of possible electromagnetic modes in a cavity, but the number of elastic modes in a finite solid is bounded. If the solid consists of N atoms, each with three degrees of freedom, the total number of modes is $3N$. An elastic wave has three possible polarizations, two transverse and one longitudinal, in contrast to the two possible polarizations of an electromagnetic wave. In a transverse elastic wave the displacement of the atoms is perpendicular to the propagation direction of the wave; in a longitudinal wave the displacement is parallel to the propagation direction. The sum of a quantity over all modes may be written as, including the factor 3:

$$\sum_{n}(\cdots) = \frac{3}{8}\int 4\pi \boldsymbol{n}^2 \mathrm{d}\boldsymbol{n}(\cdots) \tag{4.35}$$

by extension of Eq. (4.17). Here $\boldsymbol{n}$ is defined in terms of the triplet of integers $\boldsymbol{n}_x$, $\boldsymbol{n}_y$, $\boldsymbol{n}_z$, exactly as for photons. We want to find $\boldsymbol{n}_{\max}$ such that the total number of elastic modes is equal to $3N$:

$$\frac{3}{8}\int_0^{\boldsymbol{n}_{\max}} 4\pi \boldsymbol{n}^2 \mathrm{d}\boldsymbol{n} = 3N \tag{4.36}$$

In the photon problem there was no corresponding limitation on the total number of modes. It is customary to write $\boldsymbol{n}_{\mathrm{D}}$, after Debye, for $\boldsymbol{n}_{\max}$. Then Eq. (4.36) becomes

$$\frac{1}{2}\pi\,\boldsymbol{n}_{\mathrm{D}}{}^3 = 3N, \quad \boldsymbol{n}_{\mathrm{D}} = (6N/\pi)^{1/3} \tag{4.37}$$

The thermal energy of the phonons is, from Eq. (4.16),

$$U = \sum\langle \varepsilon_n\rangle = \sum\langle s_n\rangle \hbar\omega_n = \sum \frac{\hbar\omega_n}{\exp(\hbar\omega_n/\tau) - 1} \tag{4.38}$$

or, by Eq. (4.35) and Eq. (4.37),

$$U = \frac{3\pi}{2}\int_0^{np} \mathrm{d}\boldsymbol{n}\boldsymbol{n}^2 \frac{\hbar\omega_n}{\exp(\hbar\omega_n/\tau) - 1} \tag{4.39}$$

By analogy with the evaluation of Eq. (4.19), with the velocity of sound v written in place of the velocity of light c,

$$U = (3\pi^2\hbar v/2L)\,(\tau L/\pi\hbar r)^4\int_0^{xp} \mathrm{d}x \frac{x^3}{\exp x - 1} \tag{4.40}$$

where $x \equiv \pi\hbar \mathrm{v}\boldsymbol{n}/L\tau$. For L^3 we write the volume V. Here, with Eq. (4.37) the upper limit of integration is

$$x_{\mathrm{D}} = \pi\hbar v\boldsymbol{n}_{\mathrm{D}}/\mathrm{L}\tau = \hbar\mathrm{r}\,(6\pi^2\mathrm{N}/\mathrm{V})^{1/3}/\tau \tag{4.41}$$

usually written as

$$x_{\mathrm{D}} = \theta/T = k_{\mathrm{B}}\theta/\tau \tag{4.42}$$

where θ is called the Debyc temperature:

$$\theta=(\hbar v/k_B)(6\pi^2 N/V)^{1/3} \tag{4.43}$$

The result of Eq. (4.40) for the energy is of special interest at low temperatures such that $T\ll\theta$. Here the limit x_D on the integral is much larger than unity, and x_D may be replaced by infinity. We note from Figure 4.4 that there is little contribution to the integrand out beyond $x=10$. For the definite integral we have

$$\int_0^x dx \frac{x^3}{\exp x-1}=\frac{\pi^4}{15} \tag{4.44}$$

as earlier. Thus the energy in the low temperature limit is

$$U(T)\approx\frac{3\pi^4 N\tau^4}{5(k_B\theta)^3}\approx\frac{3\pi^4 Nk_B T^4}{5\theta^3} \tag{4.45}$$

proportional to T^4. The heat capacity is, for $\tau\ll k_B\theta$ or $T\ll\theta$

$$C_V=\left(\frac{\partial U}{\partial\tau}\right)_V=\frac{12\pi^4 N}{5}\left(\frac{\tau}{k_B\theta}\right)^3 \tag{4.46a}$$

$$C_V=\left(\frac{\partial U}{\partial T}\right)_V=\frac{12\pi^4 Nk_B}{5}\left(\frac{T}{\theta}\right)^3 \tag{4.46b}$$

This result is known as the Debye T^3 law. Experimental results for argon are plotted in Figure 4.10. The calculated variation of C_V versus $\frac{T}{\theta}$ is plotted in Figure 4.11. Several related thermodynamic functions for a Debye solid are plotted in Figure 4.12.

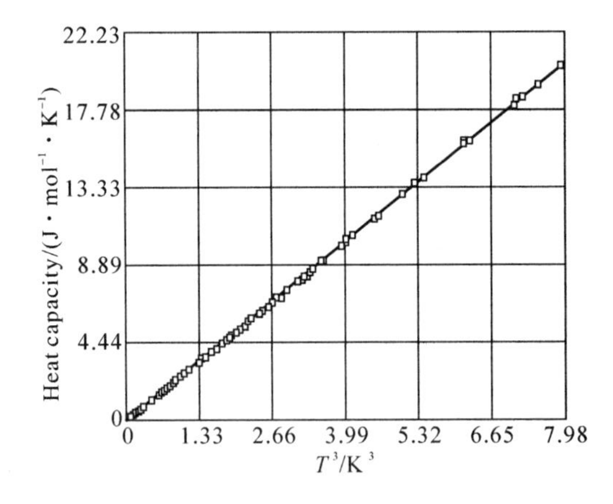

Figure 4.10 Low temperature heat capacity of solid argon, plotted against T^3 to show the excellent agreement with the Debye T^3 law

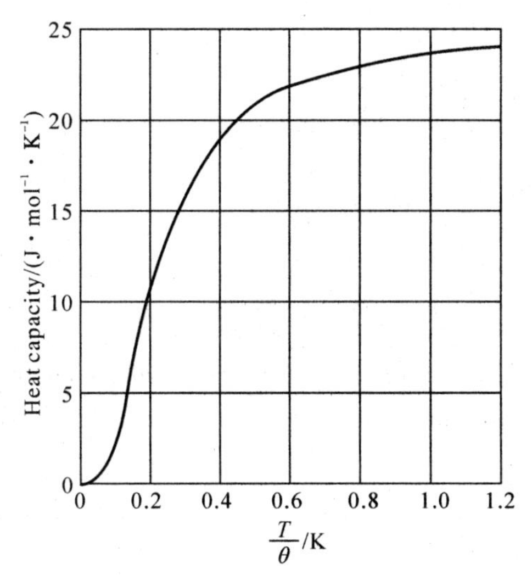

Figure 4.11 Heat capacity C_V of a solid according to the Debye approximation

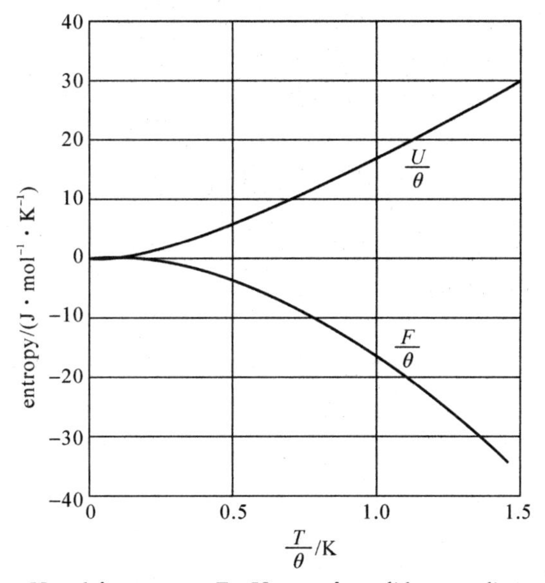

Figure 4.12 Energy U and free energy $F \equiv U - \tau\sigma$ of a solid, according to the Debye theory. The Debye temperature of the solid is θ

4.5 Summary

(1) The Planck distribution function is

$$\langle S \rangle = \frac{1}{\exp(\hbar\omega/\tau) - 1}$$

for the thermal average number of photons in a cavity mode of frequency ω.

(2) The Stefan-Boltzmann law is

$$\frac{U}{V} = \frac{\pi^2}{15\,\hbar^3 c^3}\tau^4$$

for the radiant energy density in a cavity at temperature τ.

(3) The Planck radiation law is

$$u_\omega = \frac{\hbar}{\pi^2 c^3}\frac{\omega^3}{\exp(\hbar\omega/\tau) - 1}$$

for the radiation energy per unit volume per unit range of frequency.

(4) The flux density of radiant energy is $J_U = \sigma_B T^4$ where σ_B is the Stefan-Boltzmann constant.

(5) The Debye low temperature limit of the heat capacity of a dielectric solid is, in conventional units,

$$C_V = \frac{12\pi^4 N k_B}{5}\left(\frac{T}{\theta}\right)^3$$

where the Debye temperature

$$\theta \equiv (\hbar v/k_B)\,(6\pi^2 N/V)^{1/3}$$

4.6 Problems

1. Number of thermal photons

Show that the number of photons $\sum\langle S_n\rangle$ in equilibrium at temperature τ in a cavity of volume V is

$$N = 2.404\pi^{-2} V\,(\tau/\hbar c)^3 \qquad (4.47)$$

From Eq. (4.23) the entropy is $\sigma = (4\pi^2 V/45)(\tau/\hbar c)^3$, whence $\sigma/N \approx 3.602$. It is believed that the total number of photons in the universe is 10^8 larger than the total number of nucleons (protons, neutrons). Because both entropies are of the order of the respective number of particles (see Eq. 3.76), the photons provide the dominant contribution to the entropy of the universe, although the particles dominate the total energy. We believe that the entropy of the photons is essentially constant, so that the entropy of the universe is approximately constant with time.

2. Surface temperature of the Sun

The value of the total radiant energy flux density at the Earth from the Sun normal to the incident rays is called the solar constant of the Earth. The observed value integrated over all emission wavelengths and referred to the mean Earth-Sun distance is

$$\text{solar constant} = 0.136\ \text{J} \cdot \text{s}^{-1} \cdot \text{cm}^{-2} \tag{4.48}$$

(1)Show that the total rate of energy generation of the Sun is 4×10^{26} J · s^{-1}.

(2)From this result and the Stefan-Boltzmann constant $\sigma_B = 5.67\times10^{-12}$ J · s^{-1} · cm^{-2} · K^{-4}, show that the effective temperature of the surface of the Sun treated as a black body is $T\approx 6{,}000$ K. Take the distance of the Earth from the Sun as 1.5×10^{13} cm and the radius of the Sun as 7×10^{10} cm.

3. Average temperature of the interior of the Sun

(1)Estimate by a dimensional argument or otherwise the order of magnitude of the gravitational self-energy of the Sun, with $M_{\odot} = 2\times10^{33}$ g and $R_{\odot} = 7\times10^{10}$ cm. The gravitational constant G is 6.6×10^{-8} dyne · cm^2 · g^{-2}. The self-energy will be negative referred to atoms at rest at infinite separation.

(2)Assume that the total thermal kinetic energy of the atoms in the Sun is equal to $-1/2$ times the gravitational energy. This is the result of the virial theorem of mechanics. Estimate the average temperature of the Sun. Take the number of particles as 1×10^{57}. This estimate gives somewhat too low a temperature, because the density of the Sun is far from uniform. "The range in central temperature for different stars, excluding only those composed of degenerate matter for which the law of perfect gases does not hold (white dwarfs) and those which have excessively small average densities (giants and supergiants), is between 1.5×10^7 and 3.0×10^7 degrees." (O. Struve, B. Lynds, and H. Pillans, Elementary astronomy, Oxford, 1959.)

3. Age of the Sun

Suppose 4×10^{26} J · s^{-1} is the total rate at which the Sun radiates energy at the present time.

(1)Find the total energy of the Sun available for radiation, on the rough assumptions that the energy source is the conversion of hydrogen (atomic weight 1.007,8) to helium (atomic weight 4.002,6) and that the reaction stops when 10 percent of the original hydrogen has been converted to helium. Use the Einstein relation $E=(\Delta M)c^2$.

(2)Use (1) to estimate the life expectancy of the Sun. It is believed that the age of the universe is about 10×10^9 years(A good discussion is given in the books by Peebles and by Weinberg, cited in the general references.).

5. Surface temperature of the Earth

Calculate the temperature of the surface of the Earth on the assumption that as a black body in thermal equilibrium it reradiates as much thermal radiation as it receives from the Sun. Assume also that the surface of the Earth is at a constant temperature over the day-night cycle. Use $T_{\odot}=5,800$ K; $R_{\odot}=7\times10^{10}$ cm; and the Earth-Sun distance of 1.5×10^{13} cm.

6. Pressure of thermal radiation

Show for a photon gas that:

(1)
$$p=-(\partial U/\partial V)_{\sigma}=-\sum_j s_j\hbar(\mathrm{d}\omega_j/\mathrm{d}V) \tag{4.49}$$

where s_j is the number of photons in the mode j.

(2)
$$\mathrm{d}\omega_j/\mathrm{d}V=-\omega_j/3V \tag{4.50}$$

(3)
$$p=U/3V \tag{4.51}$$

Thus the radiation pressure is equal to $\frac{1}{3}\times$(energy density).

(4) Compare the pressure of thermal radiation with the kinetic pressure of a gas of H atoms at a concentration of 1 mole · cm^{-3} characteristic of the Sun. At what temperature (roughly) are the two pressures equal? The average temperature of the Sun is believed to be near 2×10^7 K. The concentration is highly nonuniform and rises to near 100 mole · cm^{-3} at the center, where the kinetic pressure is considerably higher than the radiation pressure.

7. Free energy of a photon gas

(1) Show that the partition function of a photon gas is given by

$$Z=\prod_n[1-\exp(-\hbar\omega_n/\tau)]^{-1} \tag{4.52}$$

where the product is over the modes $\boldsymbol{n}$.

(2) The Helmholtz free energy is found directly from Eq. (4.52) as

$$F=\tau\sum_n\ln[1-\exp(-\hbar\omega_n/\tau)] \tag{4.53}$$

Transform the sum to an integral; integrate by parts to find

$$F=-\pi^2V\tau^4/45\hbar^3c^3 \tag{4.54}$$

8. Heat shields

A black (nonreflective) plane at temperature T_u is parallel to a black plane at temperature T_l. The net energy flux density in vacuum between the two planes is $J_U=\sigma_B(T_u^4-T_l^4)$, where σ_B is the Stefan-Boltzmann constant used in Eq. (4.26). A third black plane is inserted between the other two and is allowed to come to a steady state temperature T_m. Find T_m in terms of T_u and T_l, and show that the net energy flux density is cut in half because of the presence of this plane. This is the principle of the heat shield and is widely used to reduce radiant heat transfer. Comment: The result for N independent heat shields floating in temperature between the planes T_u and T_l is that the net energy flux density is

$$J_U = \sigma_B (T_u^4 - T_l^4)/(N+1)$$

9. Photon gas in one dimension

Consider a transmission line of length L on which electromagnetic waves satisfy the one-dimensional wave equation $u^2 \, \partial^2 E/\partial x^2 = \partial^2 E/\partial t^2$ where E is an electric field component. Find the heat capacity of the photons on the line, when in thermal equilibrium at temperature τ. The enumeration of modes proceeds in the usual way for one dimension: take the solutions as standing waves with zero amplitude at each end of the line.

10. Heat capacity of intergalactic space

Intergalactic space is believed to be occupied by hydrogen atoms in a concentration $\approx$ 1 atom • m^{-3}. The space is also occupied by thermal radiation at 2.9 K, from the Primitive Fireball. Show that the ratio of the heat capacity of matter to that of radiation is $\sim 10^{-9}$.

11. Heat capacity ofsolids in high temperature limit

Show that in the limit $T \gg \theta$ the heat capacity of a solid goes towards the limit $C_V \to 3Nk_B$, in conventional units. To obtain higher accuracy when T is only moderately larger than θ, the heat capacity can be expanded as a power series in $1/T$, of the form

$$C_V = 3Nk_B \left[1 - \sum_n a_n / T^n \right] \tag{4.55}$$

Determine the first nonvanishing term in the sum. Check your result by inserting $T=\theta$.

12. Heat capacity of photons and phonons

Consider a dielectric solid with a Debye temperature equal to 100 K and with 10^{22} atoms • cm^{-3}. Estimate the temperature at which the photon contribution to the heat capacity would be equal to the phonon contribution evaluated at 1 K.

13. Energy fluctuations in asolid at low temperatures

Consider a solid of N atoms in the temperature region in which the Debye T^3 law is valid. The solid is in thermal contact with a heat reservoir. Use the results on energy fluctuations from Chapter 3 to show that the root mean square fractional energy fluctuation φ, is given by

$$\varphi^2 = \langle(\varepsilon - \langle\varepsilon\rangle)^2\rangle / \langle\varepsilon\rangle^2 \approx \frac{0.07}{N}\left(\frac{\theta}{T}\right)^3 \quad (4.56)$$

Suppose that $T=10^{-2}$ K; $\theta=200$ K; and $N\approx10^{15}$ for a particle 0.01 cm on a side; then $\varphi\approx0.02$. At 10^5 K the fractional fluctuation in energy is of the order of unity for a dielectric particle of volume 1 cm^3.

14. Heat capacity of liquid ^{4}He at low temperatures

The velocity of longitudinal sound waves in liquid ^{4}He at temperatures below 0.6 K is 2.383×10^4 cm/s. There are no transverse sound waves in the liquid. The density is 0.145 g/cm^3.

(1) Calculate the Debye temperature.

(2) Calculate the heat capacity per gram on the Debye theory and compare with the experimental value $C_V=0.020,4T^3$, in $J \cdot g^{-1} \cdot K^{-1}$. The T^3 dependence of the experimental value suggests that phonons are the most important excitations in liquid ^{4}He below 0.6 K. Note that the experimental value has been expressed per gram of liquid. The experiments are due to J. Wiebes, C. G. Niels-Hakkenberg, and H. C. Kramers, Physica 32, 625 (1957).

15. Angular distribution of radiant energy flux

(1) Show that the spectral density of the radiant energy flux that arrives in the solid angle $d\Omega$ is $cu_\omega \cos\theta\, d\Omega/4\pi$, where θ is the angle the normal to the unit area makes with the incident ray, and u_ω, is the energy density per unit frequency range.

(2) Show that the sum of this quantity over all incident rays is $\frac{1}{4}cu_\omega$.

16. Image of a radiant object

Let a lens image the hole in a cavity of area A_H on a black object of area A_0. Use an equilibrium argument to relate the product $A_H\Omega_H$ to $A_0\Omega_0$ where Ω_H and Ω_0 are the solid angles subtended by the lens as viewed from the hole and from the object. This general property of focusing systems is easily derived from geometrical optics. It is also true when diffraction is important. Make the approximation that all rays are nearly parallel (all axial angles small).

17. Entropy and occupancy

We argued in this chapter that the entropy of the cosmic black body radiation has not changed with time because the number of photons in each mode has not changed with time, although the frequency of each mode has decreased as the wavelength has increased with the expansion of the universe. Establish the implied connection between entropy and occupancy of the modes, by showing that for one mode of frequency ω the entropy is a function of the photon occupancy $\langle s \rangle$ only:

$$\sigma = \langle s+1 \rangle \ln \langle s+1 \rangle - \langle s \rangle \ln \langle s \rangle \tag{4.57}$$

It is convenient to start from the partition function.

18. Isentropic expansion of photon gas

Consider the gas of photons of the thermal equilibrium radiation in a cube of volume V at temperature τ. Let the cavity volume increase; the radiation pressure performs work during the expansion, and the temperature of the radiation will drop. From the result for the entropy we know that $\tau V^{1/3}$ is constant in such an expansion.

(1) As assume that the temperature of the cosmic black-body radiation was decoupled from the temperature of the matter when both were at 3,000 K. What was the radius of the universe at that time, compared to now? If the radius has increased linearly with time, at what fraction of the present age of the universe did the decoupling take place?

(2) Show that the work done by the photons during the expansion is

$$W = (\pi^2/15\hbar^3 c^3) V_i \tau_i^3 (\tau_i - \tau_f)$$

The subscripts i and f refer to the initial and final states.

19. Reflective heat shield and Kirchhoff's law

Consider a plane sheet of material of absorptivity a, emissivity e, and reflectivity $r=1-a$. Let the sheet be suspended between and parallel with two black sheets maintained at temperatures τ_u and τ_l. Show that the net flux density of thermal radiation between the black sheets is $(1-r)$ times the flux density when the intermediate sheet is also black as in Problem 8, which means with $a=e=1$; $r=0$. Liquid helium dewars are often insulated by many, perhaps 100, layers of an aluminized Mylar film called Superinsulation.

4.7 Supplement: Greenhouse Effect

The Greenhouse Effect describes the warming of the surface of the Earth caused by the interposition of an infrared absorbent layer of water, as vapor and in clouds, and of carbon dioxide in the atmosphere between the Sun and the Earth. The water may contribute as much 90 percent of the warming effect.

Absent such a layer, the temperature of thesurface of the Earth is determined primarily by the requirement of energy balance between the flux of solar radiation incident on the Earth and the flux of reradiation from the Earth; the reradiation flux is proportional to the fourth power of the temperature of the Earth, as in Eq. (4.26). This energy balance leads to the result $T_E = (R_S/2D_{SE})^{1/2} T_S$, where T_E is the temperature of the Earth and T_S is that of the Sun; here R_S is the radius of the Sun and D_{SE} is the Sun-Earth distance.

The result of that problem is $T_E = 280$ K, assuming $T_S = 5{,}800$ K. The Sun is much hotter than the Earth, but the geometry (the small solid angle subtended by the Sun) reduces the solar flux density incident at the Earth by a factor of roughly $(1/20)^4$.

We assume as an examplethat the atmosphere is a perfect greenhouse, defined as an absorbent layer that transmits all of the visible radiation that falls on it from the Sun, but absorbs and re-emits all the radiation (which lies in the infrared), from the surface of the Earth. We may idealize the problem by neglecting the absorption by the layer of the infrared portion of the incident solar radiation, because the solar spectrum lies almost entirely at higher frequencies, as evident from Figure 4.4. The layer will emit energy flux I_L up and I_L down; the upward flux will balance the solar flux I_S, so that $I_L = I_S$ net downward flux will be the sum of the solar flux I_S and the flux I_L down from the layer. The latter increases the net thermal flux incident at the surface of the Earth. Thus

$$I_{ES} = I_L + I_S = 2I_S \tag{4.58}$$

where I_{ES} is the thermal flux from the Earth in the presence of the perfect greenhouse effect. Because the thermal flux varies as T^4, the new temperature of the surface of the Earth is

$$T_{ES} = 2^{1/4} T_E \approx (1.19)280\ \text{K} \approx 333\ \text{K} \tag{4.59}$$

so that the greenhouse warming of the Earth is 333 K−280 K=53 K for this extreme example.

Chapter 5 Chemical Potential and Gibbs Distribution

We considered in Chapter 2 the properties of two systems in thermal contact, and we were led naturally to the definition of the temperature. If the two systems have the same temperature there is no net energy flow between them. If the temperatures of two systems are different energy will flow from the system with the higher temperature to the system with the lower temperature.

Now consider systems that can exchange particles as well as energy. Such systems are said to be in diffusive (and thermal) contact: molecules can move from one system to the other by diffusion through a permeable interface. Two systems are in equilibrium with respect to particle exchange when the net particle flow is zero.

The chemical potential governs the flow of particles between the systems just as the temperature governs the flow of energy. If two systems with a single chemical species are at the same temperature and have the same value of the chemical potential there will be no net particle flow and no net energy flow between them. If the chemical potentials of the two systems are different particles will flow from the system at the higher chemical potential to the system at the lower chemical potential. As an example the chemical potential of electrons at one terminal of a storage battery is higher than at the other terminal. When the terminals are connected by a wire electrons will flow in the wire from high to low chemical potential.

Consider the establishment of diffusive equilibrium between two systems R_1 and R_2 that are in thermal and diffusive contact. We maintain τ constant by placing both systems in thermal contact (see Figure 5.1) with a large reservoir R. We found earlier that for a single system R in thermal equilibrium with a reservoir R, the Helmholtz free energy of R will assume the minimum value compatible with the common temperature τ and with other restraints on the system such as the volume and the number ofparticles. This result applies equally to the combined R_1+R_2 in equilibrium with. In diffusive equilibrium between R_1 and R_2 the particle distribution N_1, N_2 between the systems makes the total Helmholtz free energy

$$F = F_1 + F_2 = U_1 + U_2 - \tau(\sigma_1 + \sigma_2) \tag{5.1}$$

a minimum, subject to $N=N_1+N_2=$constant. Because N is constant, the Helmholtz free energy of the combined system is a minimum with respect to variations. $\delta N_1=-\delta N_2$. At the minimum,

$$dF = (\partial F_1/\partial N_1)_\tau dN_1 + (\partial F_2/\partial N_2)_\tau dN_2 = 0 \tag{5.2}$$

with V_1, V_2, also held constant. With $dN_1 = -dN_2$, we have

$$dF = [(\partial F_1/\partial N_1)_\tau - (\partial F_2/\partial N_2)_\tau]dN_1 = 0 \tag{5.3}$$

so that at equilibrium

$$(\partial F_1/\partial N_1)_\tau = (\partial F_2/\partial N_2)_\tau \tag{5.4}$$

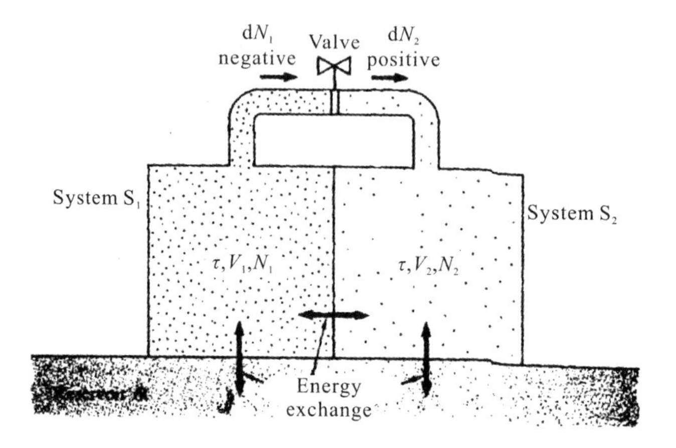

Figure. 5.1 Example of two systems, R_1 and R_2, in thermal contact with each other and with a large reservoir R, forming a closed total system

5.1 Definition of Chemical Potential

We define the chemical potential as

$$\mu(\tau, V, N) = \left(\frac{\partial F}{\partial N}\right)_{\tau, V} \tag{5.5}$$

where μ is the Greek letter mu. Then

$$\mu_1 = \mu_2$$

Expresses the condition for diffusive equilibrium. If $\mu_1 > \mu_2$, we see from Eq. (5.3) that dF will be negative when dN_1 is negative: When particles are transferred from R_1 to R_2, the value of dN_1 is negative, and dN_2 is positive. Thus the free energy decreases as particles flow from R_1 to R_2; that is, particles flow from the system of high chemical potential to the system of low chemical potential. The strict definition of μ is in terms of a difference and not a derivative because particles are not divisible:

$$\mu(\tau, V, N) \equiv F(\tau, V, N) - F(\tau, V, N-1) \tag{5.6}$$

The chemical potential regulates the particle transfer between systems in contact and it is fully as important as the temperature, which regulates the energy transfer. Two systems that can exchange both energy and particles are in combined thermal and diffusive equilibrium when their temperatures and chemical potentials arc equal:

$$\tau_1 = \tau_2/(\mu_1 - \mu_2)$$

A difference in chemical potential acts as a driving force for the transfer of particles just as a difference in temperature acts as a driving force for the transfer of energy.

If several chemical species are present, each has its own chemical potential.

For species j

$$\mu_j = (\partial F/\partial N_j)_{\tau,V,N_1,N_2,\cdots} \tag{5.7}$$

where in the differentiation the numbers of all particles are held constant except for the species j.

Example: Chemical potential of the ideal gas

In Eq. (3.70) we showed that the free energy of the monatomic ideal gas is

$$F = -\tau[N \ln Z_1 - \ln N!] \tag{5.8}$$

where

$$Z_1 = n_Q V = (M\tau/2\pi\hbar^2)^{3/2} V \tag{5.9}$$

is the partition function for a single particle. From Eq. (5.8),

$$\mu = (\partial F/\partial N)_{\tau,V} = -\tau\left(\ln Z_1 - \frac{d}{dN}\ln N!\right) \tag{5.10}$$

If we use the Stirling approximation for $N!$ and assume that we can differentiate the factorial, we find

$$\frac{d}{dN}\ln N! = \frac{d}{dN}\left[\ln\sqrt{2\pi} + \left(N+\frac{1}{2}\right)\ln N - N\right] =$$

$$\ln N + \left(N+\frac{1}{2}\right)\frac{1}{N} - 1 = \ln N + \frac{1}{2N} \tag{5.11}$$

which approaches $\ln N$ for large values of N. Hence the chemical potential of the ideal gas is

$$\mu = -\tau[\ln Z_1 - \ln N] = \tau \ln(N/Z_1)$$

or by Eq. (5.9),

$$\mu = \tau \ln(n/n_Q) \tag{5.12a}$$

where $n=N/V$ is the concentration of particles and $n_Q=(M\tau/2\pi h^2)^{3/2}$ is the quantum concentration defined by Eq. (3.63).

If we use $\mu=F(N)-F(N-1)$ from Eq. (5.6) as the definition of μ we do not need to use the Strling approximation. From Eq. (5.8) we obtain $\mu=-\tau[\ln Z_1 - \ln N]$ which agrees with Eq. (5.12a). The result depends on the concentration of particles, not on their total number or on the system volume separately. By use of the ideal gas law $p=n\tau$ we can write Eq. (5.12a) as

$$\mu = \tau\ln(p/\tau n_Q) \tag{5.12b}$$

The chemical potential increases as the concentration of particles increases. This is what we expect intuitively: particles flow from higher to lower chemical potential from higher to lower concentration. Figure 5.2 shows the dependence on concentration of an ideal gas composed of electrons or of helium atoms for two temperatures the boiling temperature of liquid helium at atmospheric pressure, 4.2 K and room temperature, 300 K. Atomic and molecular gases always have negative chemical potentials under physically realizable conditions: at classical concentrations such that $n/n_Q \ll 1$, we see from Eq. (5.12b) that μ is negative.

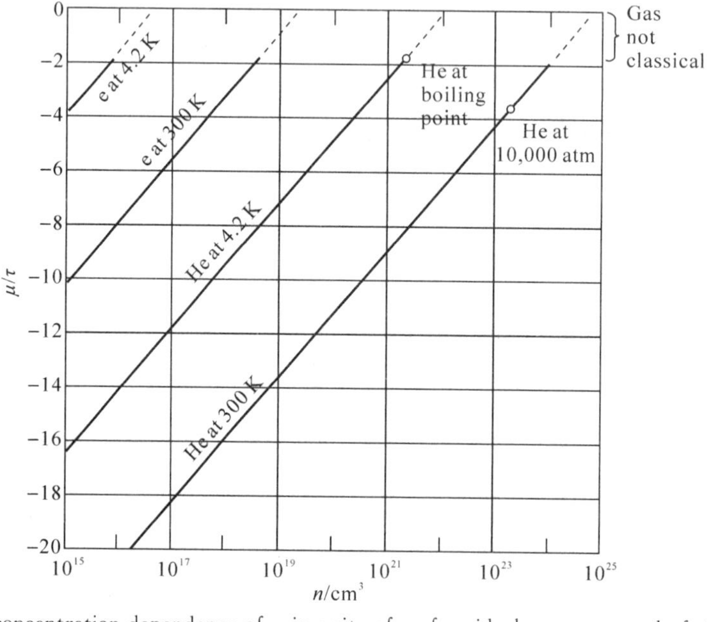

Figure 5.2 The concentration dependence of μ in units of τ of an ideal gas composed of electrons or helium atoms at 4.2 K and 300 K to be in the classical regime with $n \ll n_Q$, a gas must have a value of μ at least τ

5.2 Internal and Total Chemical Potential

The best way to understand the chemical potential is to discuss diffusive equilibrium in the presence of a potential step that acts on the particles. This problem has wide applicalion and includes the semiconductor P-N junction. We again consider two systems R_1 and R_2, at the same temperature and capable of exchanging particles but not yet in diffusive equilibrium. We assume that initially $\mu_2 > \mu_1$ and we denote the initial non-equilibrium chemical potential difference by $\Delta\mu(\text{initial}) = \mu_2 - \mu_1$. Now let a difference in potential energy be established between the two systems such that the potential energy of each particle in system R_1 is raised by exactly $\Delta\mu$(initial) above its initial value. If the particles carry a charge q, one simple way to establish this potential step is to apply between the two systems a voltage. ΔV such that

$$q\Delta V = q(V_2 - V_1) = \Delta\mu(\text{initial}) \tag{5.13}$$

with the polarity shown in Figure 5.3. A difference in gravitational potential also can serve as a potential difference: when we raise a system of particles each of mass M by the height h, we establish a potential difference Mgh, where g is the gravitational acceleration.

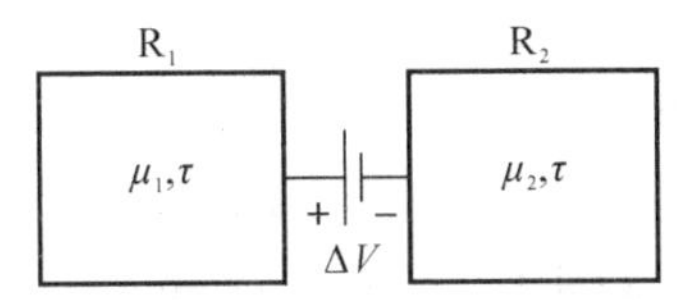

Figure 5.3 A potential step between two systems of charged particles can be established by applying a voltage between the systems

Once a potential step is present the polential energy of the particles produced by this step is included in the energy U and inthe free energy F of the system. If in Figure 5.3 we keep the free energy of system R_2 fixed. The step raises the free energy of R_1 by $N_1\Delta\mu$(initial)$=N_1 q\Delta V$ relative to its initial value. In the language of energy states to the energy of each state of R_1 the potential energy $N_1\Delta\mu$(initial) has been added. The insertion of the potential barrier specified by Eq. (5.13) raises the chemical potential of R_1 by $\Delta\mu$(initial) make the final chemical potential of R_1 equal to that of R_2:

$$\mu_1(\text{final}) = \mu_1(\text{initial}) + [\mu_2(\text{initial}) - \mu_1(\text{initial})] = \mu_2(\text{initial}) = \mu_2(\text{final}) \tag{5.14}$$

When the barrier was inserted, μ_2 was held fixed. Thus the barrier $q\Delta V=\mu_2(\text{initial})-\mu_1(\text{initial})$ brings the two systems into diffusive equilibrium.

The chemical potential is equivalent to a true potential energy: the difference in chemical potential between two systems is equal to the potential barrier that will bring the two systems into diffusive equilibrium.

This statement gives us a feeling for the physical effect of the chemical potential and it forms the basis for the measurement of chemical potential differences between two systems. To measure $\mu_2-\mu_1$, we establish a potential step between two systems that can transfer particles and we determine the step height at which the net particle transfer vanishes.

Only differences of chemical potential have a physical meaning. The absolute value of the chemical potential depends on the zero of the potential energy scale. The ideal gas result of Eq. (5.12) depends on the choice of the zero of energy of a free particle as equal to the zero of the kinetic energy.

When external potential steps are present we can express the total chemical potential of a system as the sum of two parts:

$$\mu = \mu_{tot} = \mu_{ext} + \mu_{int} \tag{5.15}$$

Here μ_{ext} is the potential energy per particle in the external potential, and μ_{int} is the internal chemical potential defined as the chemical potential that would be present if the external potential were zero. The term μ_{ext} may be mechanical, electrical, magnetic, gravitational, etc. in origin. The equilibrium condition $\mu_2 = \mu_1$ can be expressed as

$$\Delta\mu_{ext} = -\Delta\mu_{int} \tag{5.16}$$

Unfortunately, the distinction between external and internal chemical potential sometimes is not made in the literature. Some writers particularly those working with charged particles in the fields of electrochemistry and of semiconductors often mean the internal chemical potential when they use the words chemical potential without a further qualifier.

The total chemical potential may be called the electrochemical potential if the potential barriers of interest are electrostatic. Although the term electro-chemical potentialis clear and unambiguous, we shall use "total chemical potential". The use of "chemical potential" without an adjective should be avoided in situations in which any confusion about its meaning could occur.

Example: Variation barometric pressure with altitude

The simplest example of the diffusive equilibrium between systems in different external potentials is the equilibrium between layers at different heights of the Earth's atmosphere, assumed to be isothermal. The real atmosphere is in imperfect equilibrium: it is constantly upset by meteorological processes, both in the form of macroscopic air movements and of strong temperature gradients from cloud formation. and because of heat input from the ground. We may make an approximate model of the atmosphere by treating the different air layers as systems of ideal gases inthermal and diffusive equilibrium with each other, in different external potentials (see Figure 5.4). If we place the zero of the potential energy at ground level the potential energy per molecule at height h is Mgh, where M is the particle mass and g the gravitational acceleration. The internal chemical potential of the particles is given by Eq. (5.12). The total chemical potential is

$$\mu = \tau \ln(n/n_Q) + Mgh \tag{5.17}$$

In equilibrium this must be independent of the height. Thus

$$\tau \ln[n(h)/n_Q] + Mgh = \tau \ln[n(0)/n_Q]$$

and the concentration $n(h)$ at height h satisfies

$$n(h) = n(0)\exp(-Mgh/\tau) \tag{5.18}$$

The pressure of an ideal gas is proportional to the concentration: therefore the pressure at altitude h is

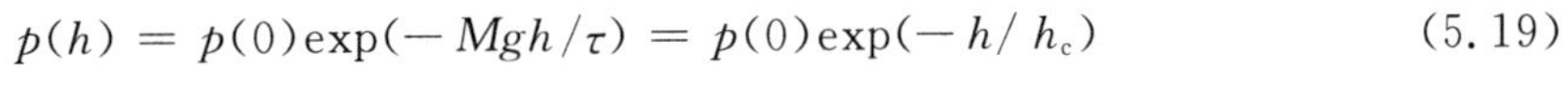

$$p(h) = p(0)\exp(-Mgh/\tau) = p(0)\exp(-h/h_c) \tag{5.19}$$

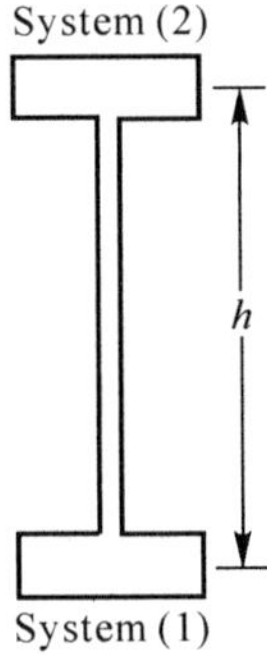

Figure 5.4 A model of the variation of atmospheric pressure with altitude: two volumes of gas at different heights in a uniform gravitational field, in thermal and diffusive contact

This is the barometric pressure equation. It gives the dependence of the pressure on altitude in an isothermal atmosphere of a single chemical species. At the characteristic height $h_c = \tau/Mg$ the atmospheric pressure decreases by the fraction $e^{-1} \approx 0.37$. To estimate the characteristic height consider an isothermal atmosphere composed of nitrogen molecules with a molecular weight of 28. The mass of an N_2 molecule is 48×10^{-24} g/m. At a temperature of 290 K the value of $\tau \equiv k_B T$ is 4.0×10^{-14} erg. With $g = 980$ cm/s^2 the characteristic height h is 8.5 km, approximately 5 miles. Lighter molecules, H_2 and He, will extend farther up, but these have largely escaped from the atmosphere.

Because the Earth's atmosphere is not accurately isothermal $n(h)$ has a more complicated behavior. Figure 5.5 is a logarithmic plot of pressure data between 10 and 40 kilometers, taken on rocket flights. The data points fall near a straight line suggesting roughly isothermal behavior. The straight line connecting the data points of Figure 5.5 spans a pressure range $p(h_2) : p(h_1) = 1{,}000 : 1$ over an altitude range from $h_1 = 2$ km to $h_2 = 48$ km. Now, from Eq. (5.19)

$$\ln \frac{p(h_1)}{p(h_2)} = \frac{Mg}{\tau}(h_2 - h_1) \tag{5.20}$$

so that the slope of the line is Mg/τ, which leads to $T = \tau/k_B = 227$ K. The non-intersection of the observed cue with the point $h = 0$, $p(h)/p(0) = 1$, is caused by the higher temperature at lower altitudes.

The atmosphere consists of more than one species of gas. In atomic percent the composition of dry air at sea level is 78 pct N_2, 21 pct O_2 and 0.9 pct Ar; other constituents account for less than 0.1 pct each. The water vapor content of the atmosphere may be appreciable: at $T = 300$ K (27℃), a relative humidity of 100 pct corresponds to 3.5 pct H_2O. The carbon dioxide concentration varies about a nominal value of 0.03 pct. In an ideal static isothermal atmosphere each gas would be in equilibrium with itself. The concentration of each would fall off with a separate Boltzmann factor of the form $\exp(-Mgh/\tau)$ with M the appropriate molecular mass. Because of the differences in mass, the different constituents fall off at different rates.

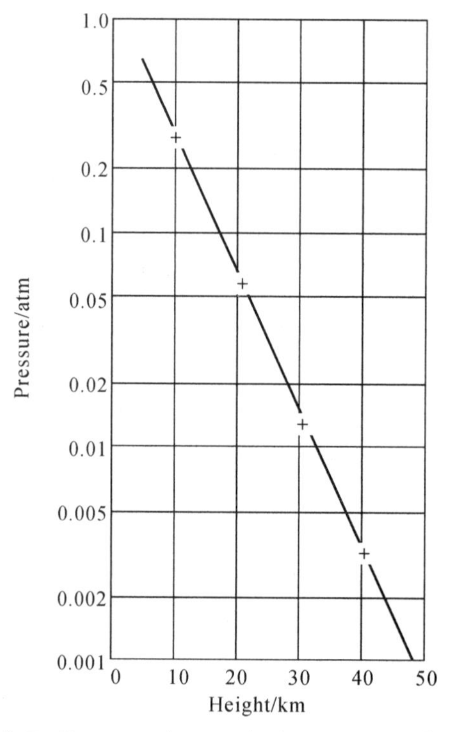

Figure 5.5 Decrease of atmospheric pressure with altitude

Example: Chemical potential of mobile magnetic particles in a magnetic field

Consider a system of N identical particles each with a magnetic moment m. For simplicity suppose each moment is directed either parallel $\uparrow$ or antiparallel $\downarrow$ to an applied magnetic field B. Then the potential energy of a $\uparrow$ particle is $-mB$, and the potential energy of a $\downarrow$ particle is $+mB$. We may treat the particles as belonging to the two distinct chemical species labelled $\uparrow$ and $\downarrow$, one with external chemical potential $\mu_{\text{ext}}(\uparrow)=-mB$ and the other with $\mu_{\text{ext}}(\downarrow)=mB$. The particles $\uparrow$ and $\downarrow$ are as distinguishable as two different isotopes of an element or as two different elements; we speak of $\uparrow$ and $\downarrow$ as distinct species in equilibrium with each other. The internal chemical potentials of the particles viewed as ideal gases with concentrations $n_\uparrow$ and $n_\downarrow$, are

$$\mu_{\text{int}}(\uparrow)=\tau\ln(n_\uparrow/n_Q),\mu_{\text{int}}(\downarrow)=\tau\ln(n_\downarrow/n_Q) \tag{5.21}$$

where $n_Q=(M\tau/2\pi\hbar^2)^{3/2}$ is the same for both species.

The total chemical potentials are

$$\mu_{tot}(\uparrow) = \tau\ln(n_{\uparrow}/n_Q) - mB \tag{5.22a}$$

$$\mu_{tot}(\downarrow) = \tau\ln(n_{\downarrow}/n_Q) + mB \tag{5.22b}$$

If the magnetic field B varies in magnitude over the volume of the system the concentration $n_{\uparrow}$ must vary over the volume in order to maintain a constant total chemical potential $\mu_{tot}(\uparrow)$ over the volume (see Figure 5.6) (The total chemical potential of a species is constant independent of position, if there is free diffusion of particles within the volume.). Because the two species m equilibrium have equal chemical potentials,

$$\mu_{tot}(\uparrow) = \text{constant} = \mu_{tot}(\downarrow) \tag{5.23}$$

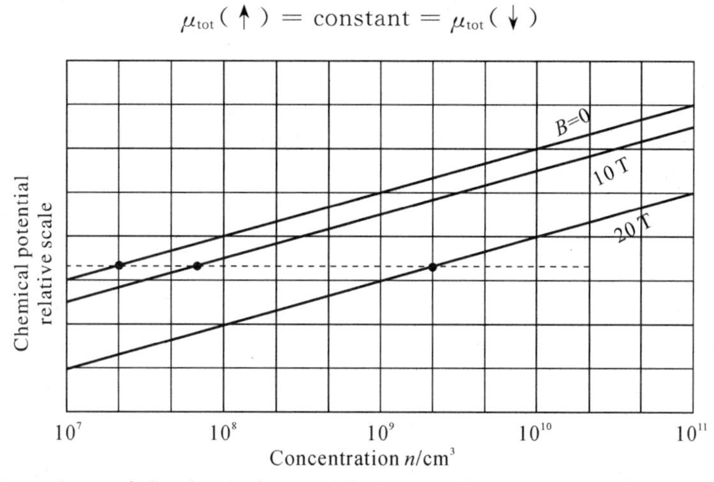

Figure 5.6 Dependence of the chemical potential of a gas of magnetic particles on the concentration at several values of the magnetic field intensity

The desired solutions of Eq. (5.22) and Eq. (5.23) are easily seen by substitution to be:

$$n_{\uparrow}(B) = \frac{1}{2}n(0)\exp(mB/\tau),\ n_{\downarrow}(B) = \frac{1}{2}n(0)\exp(-mB/\tau) \tag{5.24}$$

where $n(0)$ is the total concentration $n_{\uparrow}+n_{\downarrow}$ at a point where the field $B=0$. The total concentration at a point at magnetic field B is

$$\left.\begin{aligned} n(B) &= n_{\uparrow}+n_{\downarrow} = \frac{1}{2}n(0)[\exp(mB/\tau)+\exp(-mB/\tau)] \\ n(B) &= n(0)\cosh(mB/\tau) \approx n(0)(1+\frac{m^2B^2}{2\tau^2}+\cdots) \end{aligned}\right\} \tag{5.25}$$

The result showsthe tendency of magnetic particles to concentrate in regions of high magnetic field intensity. The functional form of the result is not limited to atoms with two magnetic orientations but is applicable to fine ferromagnetic particles in suspension in a colloidal solution. Such suspensions are used in the laboratory in the study of the magnetic flux structure of superconductors and the domain structure of ferromagnetic materials. In engineering the suspensions arc used to test for fine structural cracks in high strength steel such as turbine blades and aircraft landing gear. When these are coated with a ferromagnetic suspension and placed in a magnetic field the particle concentration becomes enhanced at the intense fields at the edges of the crack.

In the preceding discussionwe added μ_{ext} to the internal chemical potential of the particles. If the particles were ideal gas atoms, μ_{int} would be given by Eq. (5. 12). The logarithmic form for μ_{int} is not restricted to ideal gases, but is a consequence of the conditions that the particles do not interact and that their concentration is sufficiently low. Hence Eq. (5. 12) applies to macroscopic particles as wellas to atoms that satisfy these assumptions. The only difference is the value of the quantum concentration n_Q. We can therefore write

$$\mu_{int} = \tau \ln n + \text{constant} \tag{5. 26}$$

where the constant ($= -\tau \ln n_Q$) does not depend on the concentration of the particles.

Example: Batteries

One of the most vivid examples of chemical potentials and potential steps is the electrochemical battery. In the familiar lead-acid battery the negative electrode consists ofmetallic lead, Pb, and the positive electrode is a layer of reddish-brown lead oxide. PbO_2, on a Pb substrate. The electrodes are immersed in diluted sulfuric acid, H_2SO_4, which is partially ionized into H^+ ions (protons) and SO_4^{--} ions (see Figure 5. 7). It is the ions that matter.

In the discharge process both the metallic Pb of the negative electrode and the PbO_2 of the positive electrode are converted to lead sulfate, $PbSO_4$, via the two reactions:

Negative electrode:

$$Pb + SO_4^{--} \rightarrow PbSO_4 + 2e^- \tag{5. 27a}$$

Positive electrode:

$$PbO_2 + 2H^+ + H_2SO_4 + 2e^- \rightarrow PbSO_4 + 2H_2O \tag{5. 27b}$$

Because of Eq. (5. 27a) the negative electrode aαs as a sink for SO_4 ions, keeping the internal chemical potential $\mu(SO_4^{--})$ of the sulfate ions at the surface of the negative electrode lower than inside the electrolyte [see Figure 5. 7(b)]. Similarly because of Eq. (5. 27b) the positive electrode acts as a sink for H^+ ions, keeping the internal chemical potential $\mu(H^+)$ of the hydrogen ions lower at the surface of the positive electrode than inside the electrolyte. The chemical potentialgradients drive the ions towards the electrodes and they drive the electrical currents during the discharge process.

If the battery terminalsare not connected, electrons are depleted from the positive electrode and accumulate in the negative electrode thereby charging both. As a result, electrochemical potential step developat the electrode-electrolyte interfaces, steps of exactly the correct magnitude to equalize the chemical potential steps and to stop the diffusion of ions, which stops the chemical reactions from proceeding further. If an external current is permitted to flow, the reactions resume. Electron flow directly, throughthe electrolyte is negligible, because of a negligible electron concentration in the electrolyte.

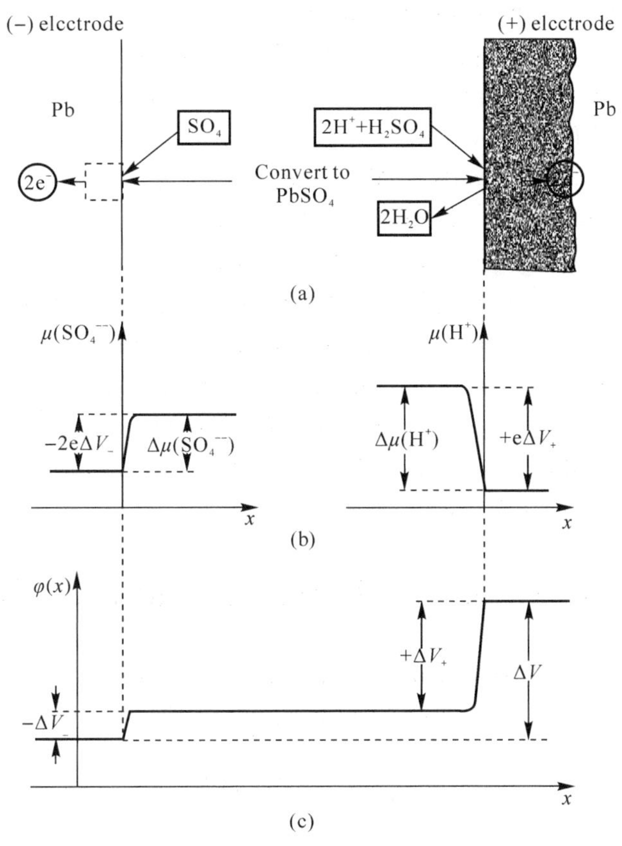

Figure 5. 7 The sulfate ions at the surface of the negative electrode

(a) The lead-acid battery consists of a Pb and a PbO_2 electrode immersed in partially ionized H_2SO_4; (b) The electrochemical potentials for SO_4^{--} and H^+ before the development of internal potential barriers that stop the diffusion and the chemical reaction; (c) The electrostatic potential $\varphi(x)$ after the formation of the barrier

During the charging process the reactions opposite to Eq. (5. 27a, b) take place, because now an external voltage is applied that generates electrostatic potential steps at the surface of the electrode of such magnitude as to reverse the sign of the (total) chemical potential gradients and hence the direction of ion flow.

We denote by ΔV and ΔV_- the differences in electrostatic potential ofthe negative and positive electrodes relative to the common electrolyte. Because the sulfate ions carry two negative charges diffusion will stop when

$$-2q\,\Delta V_- = \mu(SO_4^{--}) \tag{5.28a}$$

Diffusion of the H^+ ions will stop when

$$+q\,\Delta V_+ = \mu(H^+) \tag{5.28b}$$

The two potentials ΔV_- and ΔV_+ are called half-cell potentials or half-cell EMF's (electromotive forces): their magnitudes are known:

$$\Delta V_- = -0.4\ \text{V},\ \Delta V_+ = +1.6\ \text{V}$$

The total electrostatic potential difference developed across one full cell of the battery as required to stop the diffusion reaction, is

$$\Delta V = \Delta V_+ - \Delta V_- = 2.0\ \text{V} \tag{5.29}$$

This is the open-circuit voltage or EMF of the battery. It drives the electrons from the negative terminal to the positive terminal when the two are connected.

We have ignored free electrons in the electrolyte. The potential step tend to drive electrons from the negative electrodes into the electrolyte and from the electrolyte into the positive electrode. Such an electron current is present but the magnitude is so small as to be practically negligible, because the concentrationof electrons in the electrolyte is many orders of magnitude less than that of the ions. The only effective electron flow path is through the external connection between the electrodes.

5.3 Chemical Potential and Entropy

In Eq. (5.5) we defined the chemical potential as a derivative of the Helmholtz free energy. Here we derive an alternate relation, needed later:

$$\frac{\mu(U,V,N)}{\tau} = -\left(\frac{\partial \sigma}{\partial N}\right)_{U,V} \tag{5.30}$$

This expressesthe ratio μ/τ as a derivative of the entropy similarto the way $1/\tau$ was defined in Chapter 2.

To derive Eq. (5.30) consider the entropy as a function of the independent variables U, V, and N. The differential

$$\mathrm{d}\sigma = \left(\frac{\partial \sigma}{\partial U}\right)_{V,N}\mathrm{d}U + \left(\frac{\partial \sigma}{\partial V}\right)_{U,N}\mathrm{d}V + \left(\frac{\partial \sigma}{\partial N}\right)_{U,V}\mathrm{d}N \tag{5.31}$$

gives the differential change of the entropy for arbitrary, independent differential changes $\mathrm{d}U$, $\mathrm{d}V$, and $\mathrm{d}N$. Let $\mathrm{d}V=0$ for the processes under consideration. Further select the ratios of $\mathrm{d}\sigma$, $\mathrm{d}U$, and $\mathrm{d}N$ in such a way that the overall temperature change $\mathrm{d}\tau$ will be zero. If we denote these interdependentvalues of $\mathrm{d}\sigma$, $\mathrm{d}U$, and $\mathrm{d}N$ by $(\delta\sigma)_\tau$, $(\delta U)_\tau$ and $(\delta N)_\tau$, then $\mathrm{d}\tau=0$ when

$$(\delta\sigma)_\tau = \left(\frac{\partial \sigma}{\partial U}\right)_N \delta_\tau + \left(\frac{\partial \sigma}{\partial N}\right)_U \delta_\tau$$

After division by $(\delta N)_\tau$

$$\frac{(\delta\sigma)_\tau}{(\delta N)_\tau} = \left(\frac{\partial\sigma}{\partial U}\right)_N \frac{(\delta U)_\tau}{(\delta N)_\tau} + \left(\frac{\partial\sigma}{\partial N}\right)_U \tag{5.32}$$

The ratio $\frac{(\delta\sigma)_\tau}{(\delta N)_\tau}$ is $\left(\frac{\partial\sigma}{\partial U}\right)_N$ and $\frac{(\delta U)_\tau}{(\delta N)_\tau}$ is $\left(\frac{\partial U}{\partial N}\right)_\tau$, all at constant volume. With the original definition of $1/\tau$, We have

$$\tau\left(\frac{\partial\sigma}{\partial N}\right)_{\tau,V} = \left(\frac{\partial U}{\partial N}\right)_{\tau,V} + \tau\left(\frac{\partial\sigma}{\partial N}\right)_{U,V} \tag{5.33}$$

This expresses a derivative at constant U in terms of derivatives at constant τ. By the original definition of the chemical potential,

$$\mu \equiv \left(\frac{\partial F}{\partial N}\right)_{\tau,V} \equiv \left(\frac{\partial U}{\partial N}\right)_{\tau,V} - \tau\left(\frac{\partial\sigma}{\partial N}\right)_{\tau,V} \tag{5.34}$$

and on comparisonwith (5.33) we obtain

$$\mu = -\tau\,(\partial\sigma/\partial N)_{U,V} \tag{5.35}$$

The two expressions Eq. (5.5) and Eq. (5.35) represent two different ways to express the same quantity μ. The difference between them is the following. In Eq. (5.5), F is a function of its natural independent variables τ, V, and N, so that μ appears as a function of the same variables. In Eq. (5.31) we assumed $\sigma=\sigma(U, V, N)$, so that Eq. (5.35) yields μ as a function of U, V, N. The quantity μ is the same in both Eq. (5.5) and Eq. (5.35), but expressed in terms of different variables(see Table 5.1). The third relation for μ:

$$\mu(\sigma,V,N) = (\partial U/\partial N)_{\sigma,V} \tag{5.36}$$

Table 5.1 Summary of relations expressing the temperature τ, the pressure p, and the chemical potential μ in terms of partial derivatives of the entropy σ the energy U

	$\sigma(U, V, N)$	$U(\sigma, V, N)$	$F(\tau, V, N)$
τ	$\frac{1}{\tau}=\left(\frac{\partial\sigma}{\partial U}\right)_{V,N}$	$\tau=\left(\frac{\partial U}{\partial\sigma}\right)_{V,N}$	τ is independent variable
p	$\frac{p}{\tau}=\left(\frac{\partial\sigma}{\partial V}\right)_{U,N}$	$-p=\left(\frac{\partial U}{\partial V}\right)_{\sigma,N}$	$-p=\left(\frac{\partial F}{\partial V}\right)_{\tau,N}$
μ	$-\frac{\mu}{\tau}=\left(\frac{\partial\sigma}{\partial N}\right)_{U,V}$	$\mu=\left(\frac{\partial U}{\partial N}\right)_{\sigma,V}$	$\mu=\left(\frac{\partial F}{\partial N}\right)_{\tau,V}$

1. Thermodynamic identity

We can generalize the statement of the thermodynamic identity given in Eq. (3.34a) to include systems in which the number of particles is allowed to change. As in Eq. (5.31),

$$d\sigma = \left(\frac{\partial\sigma}{\partial U}\right)_{V,N} dU + \left(\frac{\partial\sigma}{\partial V}\right)_{U,N} dV + \left(\frac{\partial\sigma}{\partial N}\right)_{U,V} dN \tag{5.37}$$

By use of the definition Eq. (2. 26) of $1/\tau$ the relation Eq. (3. 32) for p/τ and the relation Eq. (5. 30) for $-\mu/\tau$ we write $d\sigma$ as

$$d\sigma = dU/\tau + pdV/\tau - \mu dN/\tau \tag{5.38}$$

This maybe rearranged to give

$$dU = \tau d\sigma - pdV + \mu dN \tag{5.39}$$

which is a broader statement of the thermodynamic identity than we were able to develop in Chapter 3.

5. 4 Gibbs Factor and Gibbs Sum

The Boltzmann factor derived in Chapter 3 allows us to give the ratio of the probability that a system will be in a state of energy 1∶1 to the probability the system will be in a state of energy for a system in thermal contact with a reservoir at temperature τ:

$$\frac{p(\varepsilon_1)}{p(\varepsilon_2)} = \frac{\exp(-\varepsilon_1/\tau)}{\exp(-\varepsilon_2/\tau)} \tag{5.40}$$

This is perhapsthe best known result of statistical mechanics. The Gibbs factor is the generalization of the Boltzmann factor to a system in thermal and diffusive contact with a reservoir at temperature τ and chemical potential μ. The argument retraces much of that presented in Chapter 3.

We consider a very large bodywith constant energy U_0 and constant particle number N_0. The body is composed of two parts, the very large reservoir R and the system, in thermal and diffusive contact (see Figure 5. 8). They may exchange particles and energy. The contact assures that the temperature and the chemical potential of the system are equal to those of the reservoir. When the system has N particles, the reservoir has $N_0 - N$ particles; when the system has energy ε: the reservoir has energy $U_0 - \varepsilon$. To obtain the statistical properties of the system, we make observations as before on identical copies of the system+reservoir, one copy for each accessible quantum state of the combination. What is the probability in a given observation that the system will be found to contain N particles and to be in a state s of energy R_2?

The state s is a state of a system having some specified number of particles. The energy is the energy of the state s of the N-particle system; sometimes we write only R_2, if the meaning is clear. When can we write the energy of a system having N particles in an orbital as N times the energy of one particle in the orbital? Only when interactions between the particles are neglected so that the particles may be treated as independent of each other.

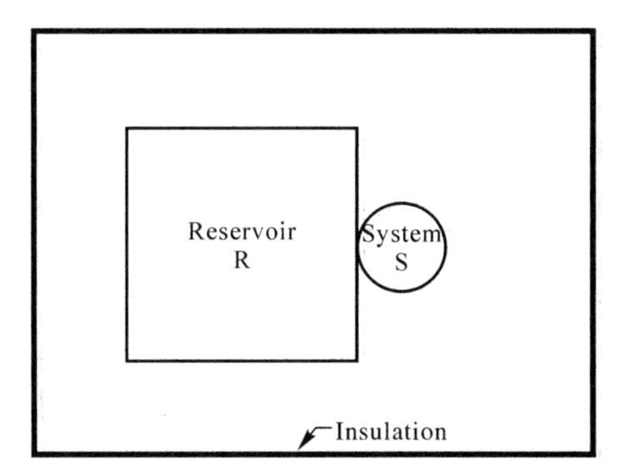

Figure 5.8　A system in thermal and diffusive contact with a large reservoir of energy and of particles

Let $P(N, R_2)$ denote the probability that the system has N particles and is in a particular state s. This probability is proportional to the number of accessible states of the reservoir when the state of the system is exactly specified. That is, if we specify the state of R, the number of accessible states of R is just the number of accessible states of R:

$$g(\mathrm{R}) = g(\mathrm{R})l \tag{5.41}$$

The factor l reminds us that we are looking at the system in a single specified state. The $g(\mathrm{R})$ states of the reservoir have $N_0 - N$ particles and have energy $U_0 - \varepsilon_s$. Because the system probability $P(N, \varepsilon_s)$ is proportional to the number of accessible states of the reservoir,

$$P(N, \varepsilon_s) \propto g(N_0 - N, U_0 - \varepsilon_s) \tag{5.42}$$

here g refers to the reservoir alone and depends on the number of particles in the reservoir and on the energy of the reservoir.

We can express Eq. (5.42) as aratio of two probabilities, one that the system is in state 1 and the other that the system is in state 2:

$$\frac{P(N_1, \varepsilon_1)}{P(N_2, \varepsilon_1)} = \frac{g(N_0 - N_1, U_0 - \varepsilon_1)}{g(N_0 - N_2, U_0 - \varepsilon_2)} \tag{5.43}$$

where g refers to the state of the reservoir. The situation is shown in Figure 5.9.

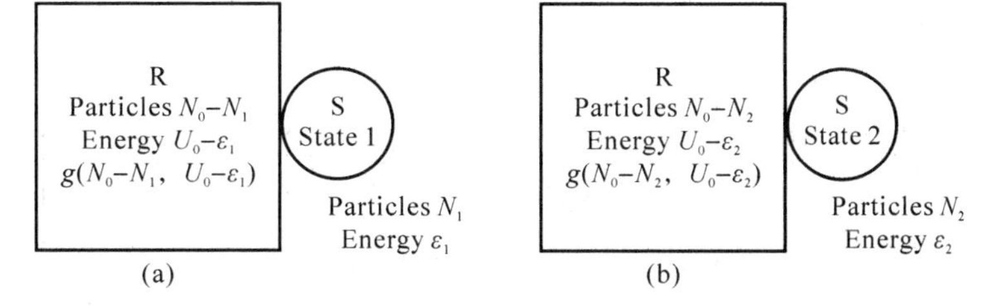

Figure 5.9　The reservoir is in thermal and diffusive contact with the system

(a) the system is in quantum state 1, and the reservoir has $g(N_0 - N_1, U_0 - \varepsilon_1)$ states accessible to it; (b) the system is in quantum state 2, and the reservoir has $g(N_0 - N_2, U_0 - \varepsilon_2)$ states accessible to it

By definitionof the entropy

$$g(N_0, U_0) \equiv \exp[\sigma(N_0, U_0)] \tag{5.44}$$

so that the probability ratioin (5.43) may be written as

$$\frac{P(N_1, \varepsilon_1)}{P(N_2, \varepsilon_2)} = \frac{\exp[\sigma(N_0 - N_1, U_0 - \varepsilon_1)]}{\exp[\sigma(N_0 - N_2, U_0 - \varepsilon_2)]} \tag{5.45}$$

or

$$\frac{P(N_1, \varepsilon_1)}{P(N_2, \varepsilon_2)} = \exp[\sigma(N_0 - N_1, U_0 - \varepsilon_1)] - \sigma(N_0 - N_2, U_0 - \varepsilon_2) = \exp(\Delta\sigma) \tag{5.46}$$

here, $\Delta\sigma$ is the entropy difference:

$$\Delta\sigma \equiv \sigma(N_0 - N_1, U_0 - \varepsilon_1) - \sigma(N_0 - N_2, U_0 - \varepsilon_2) \tag{5.47}$$

The reservoir is very large in comparison with the system and $\Delta\sigma$ may be approximated quite accurately by the first order terms in a series expansion in the two quantities N and ε that relate to the system. The entropy of the reservoir becomes

$$\sigma(N_0 - N, U_0 - \varepsilon) = \sigma(N_0, U_0) - N\left(\frac{\partial\sigma}{\partial N_0}\right)_{U_0} - \varepsilon\left(\frac{\partial\sigma}{\partial U_0}\right)_{N_0} \tag{5.48}$$

For $\Delta\sigma$ defined by Eq. (5.47) we have, to the first order in $N_1 - N_2$ and in $\varepsilon_1 - \varepsilon_2$,

$$\Delta\sigma = -(N_1 - N_2)\left(\frac{\partial\sigma}{\partial N_0}\right)_{U_0} - (\varepsilon_1 - \varepsilon_2)\left(\frac{\partial\sigma}{\partial U_0}\right)_{N_0} \tag{5.49}$$

We know that

$$-\frac{1}{\tau} \equiv \left(\frac{\partial\sigma}{\partial U_0}\right)_{N_0} \tag{5.50a}$$

by our original definition of the temperature. This is written for the reservoir, but the system will have the same temperature. Also,

$$-\frac{\mu}{\tau} \equiv \left(\frac{\partial\sigma}{\partial N_0}\right)_{U_0} \tag{5.50b}$$

by Eq. (5.30).

The entropy difference Eq. (5.49) is

$$\Delta\sigma = \frac{(N_1 - N_2)\mu}{\tau} - \frac{\varepsilon_1 - \varepsilon_2}{\tau} \tag{5.51}$$

here $\Delta\sigma$ refers to the reservoir, but $N_1, N_2, \varepsilon_1, \varepsilon_2$ refer to the system. The central result of statistical mechanics is found on combining Eq. (5.46) and Eq. (5.51):

$$\frac{P(N_1, \varepsilon_1)}{P(N_2, \varepsilon_2)} = \frac{\exp[(N_1\mu - \varepsilon_1)/\tau]}{\exp[(N_2\mu - \varepsilon_2)/\tau]} \tag{5.52}$$

The probability is the ratio of two exponential factors each of the form $\exp[(N\mu - \varepsilon)/\tau]$. A term of this form is called a Gibbs factor. The Gibbs factor is proportional to the probability that the system is in a state s of energy ε_s, and number of particles N. The result was-first given by J. W. Gibbs, who referred to it as the grand canonical distribution.

The sum of Gibbsfactors taken over all states of the system for all numbers of particles is the normalizing factor that converts relative probabilities to absolute probabilities:

$$\mathfrak{z}(\mu,\tau) = \sum_{N=0}^{\infty}\sum_{s(N)}\exp[(N\mu - \varepsilon_{s(N)})/\tau] = \sum_{ASN}\exp(N\mu - \varepsilon_{s(N)}) \tag{5.53}$$

This is called the Gibbs sum or the grand sum or the grand partition function. The sum is to be carried out over all states of the system for all numbers of particles: this defines the abbreviation ASN. We have written ε_s as $\varepsilon_{s(N)}$ to emphasize the dependence of the state on the number of particles N. That is, $\varepsilon_{s(N)}$ is the energy of the state $s(N)$ of the exact N-particle hamiltonian. The term $N=0$ must be included; if we assign its energy as zero, then the first term in $\mathfrak{z}$ will be l.

The absolute probability that the system will be found in a state N_1, ε_1 is given by the Gibbs factor divided by the Gibbs sum:

$$P(N_1, \varepsilon_1) = \frac{\exp[(N_1\mu - \varepsilon_1)/\tau]}{\mathfrak{z}} \tag{5.54}$$

This applies to a system that is at temperature τ and chemical potential μ. The ratio of any two P's is consistent with our central result of Eq. (5.52) for the Gibbs factors. Thus Eq. (5.52) gives the correct relative probabilities for the states N_1, ε_1 and N_2, ε_2. The sum of the probabilities of all states for all numbers of particles of the system is unity:

$$\sum_N\sum_s P(N, \varepsilon_s) = \sum_{ASN} P(N, \varepsilon_s) = \frac{\sum_{ASN}\exp[(N\mu - \varepsilon_{s(N)})/\tau]}{\mathfrak{z}} \tag{5.55}$$

by the definition of $\mathfrak{z}$. Thus Eq. (5.54) gives the correct absolute probability.

Average values over the systems in diffusive and thermal contact with a reservoir are easily found. If $X(N, s)$ is the value of X when the system has N particles and is in the quantum states then the thermal average of X over all N and all s is

$$\langle X\rangle = \sum_{ASN} X(N,s)P(N,\varepsilon_s) = \frac{\sum_{ASN}\exp[(N\mu - \varepsilon_{s(N)})/\tau]}{\mathfrak{z}} \tag{5.56}$$

We shall use this result to calculate thermal averages.

1. Number of particles

The number of particles in the system can vary because the system is in diffusive contact with a reservoir. The thermal average of the number of particles in the system is

$$\langle N\rangle = \frac{\sum_{ASN}\exp[(N\mu - \varepsilon_s)/\tau]}{\mathfrak{z}} \tag{5.57}$$

according to Eq. (5.56). To obtain the numerator, each term in the Gibbs sum has been multiplied by the appropriate value of N. More convenient forms of $\langle N\rangle$ can be obtained from the definition of $\mathfrak{z}$:

$$\frac{\partial \mathfrak{z}}{\partial \mu}=\frac{1}{\tau}\frac{\sum\limits_{\mathrm{ASN}}\exp[(N\mu-\varepsilon_s)/\tau]}{\mathfrak{z}} \tag{5.58}$$

whence

$$\langle N\rangle=\frac{\tau}{\mathfrak{z}}\frac{\partial \mathfrak{z}}{\partial \mu}=\tau\frac{\partial \ln \mathfrak{z}}{\partial \mu} \tag{5.59}$$

The thermal average number of particles is easily found from the Gibbs sum $\mathfrak{z}$ by direct use of Eq. (5.59). When no confusion arises, we shall write N for the thermal average $\langle N\rangle$. When we speak later of the occupancy of an orbital, f or $\langle f\rangle$ will be written interchangeably for N or $\langle N\rangle$.

We often employ the handy notation

$$\lambda\equiv\exp(\mu/\tau) \tag{5.60}$$

where λ is called the absolute activity. Here λ is the Greek letter lambda. We see from Eq. (5.12) that for an ideal gas λ is directly proportional to the concentration.

The Gibbs sum is written as

$$\mathfrak{z}=\sum_N\sum_s\lambda^N\exp(-\varepsilon_s/\tau)=\sum_{\mathrm{ASN}}\lambda^N\exp(-\varepsilon_s/\tau) \tag{5.61}$$

and the ensemble average number of particles Eq. (5.57) is

$$\langle N\rangle=\lambda\frac{\partial}{\partial\lambda}\ln\mathfrak{z} \tag{5.62}$$

This relation is useful, because inmany actual problems. We determine λ by finding the value that will make $\langle N\rangle$come out equal to the given number of particles.

2. Energy

The thermal average energy of the system is

$$U=\langle\varepsilon\rangle=\frac{\sum\limits_{\mathrm{ASN}}\varepsilon_s\exp[(N\mu-\varepsilon_s)/\tau]}{\mathfrak{z}} \tag{5.63}$$

where we have temporarily introduced the notation $\beta=1/\tau$. We shall usually write U for $\langle\varepsilon\rangle$. Observe that

$$\langle N\mu-\varepsilon\rangle=\langle N\rangle\mu-U=\frac{1}{\mathfrak{z}}\frac{\partial \mathfrak{z}}{\partial \beta}=\frac{\partial}{\partial\beta}\ln\mathfrak{z} \tag{5.64}$$

so that Eq. (5.59) and Eq. (5.63) may be combined to give

$$U=\left(\frac{\mu}{\beta}\frac{\partial}{\partial\mu}-\frac{\partial}{\partial\beta}\right)\ln\mathfrak{z}=\left[\tau\mu\frac{\partial}{\partial\mu}-\frac{\partial}{\partial(1/\tau)}\right]\ln\mathfrak{z} \tag{5.65}$$

A simpler expression that is more widely used in calculations was obtained in Chapter 3 in terms of the partition function Z.

Example: Occupancy zero or one

A red-blooded example of a system that may be occupied by zero molecules or by one molecule is the home group, which may be vacant or may be occupied by one O_2 molecule—and never by more than one O_2 molecule (see Figure 5.10). A single home group occurs in the protein myoglobin, which is responsible for the red color of meat. If ε is the energy of an adsorbed molecule of O_2 relative to O_2 at rest at infinite distance then the Gibbs sum is infinite distance, then the Gibbs sum is

$$\mathfrak{z} = 1 + \lambda \exp(-\varepsilon/\tau) \tag{5.66}$$

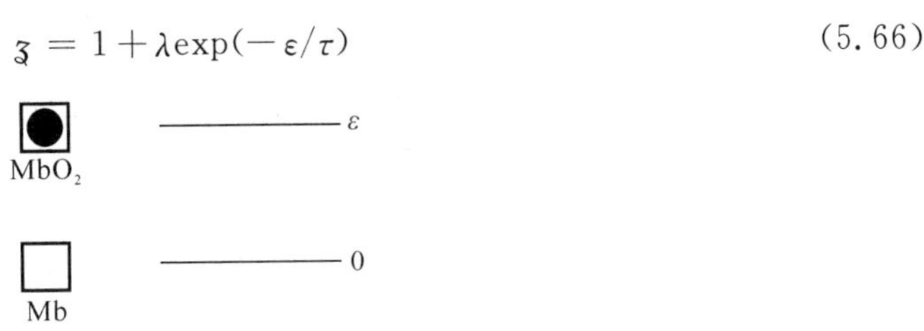

Figure 5.10 Adsorption of an O_2 by a heme, where ε is the energy of an adsorbed O_2 relative to an O_2 at infinite separation from the site

If energy must be added to remove the atom from the heme, ε will be negative. The term l in the sum arises from occupancy zero: the term $\lambda\exp(-\varepsilon/\tau)$ arises from single occupancy. These are the only possibilities. We have $Mb+O_2$ or MbO_2 present where Mb denotes myoglobin, a protein of molecular weight 17,000.

Experimental results for the fractional occupancy versus the concentration of oxygen are shown in Figure 5.11. We compare the observed oxygen saturation curves of myoglobin and hemoglobin in Figure 5.12. Hemoglobin is the oxygen-carrying component of blood. It is made up of four molecular strands each strand nearly identical with the single strand of myoglobin and each capable of binding a single oxygen molecule. Historically the classic work on the adsorption of oxygen by hemoglobin was done by Christian Bohr, the father of Niels Bohr. The oxygen saturation curve for hemoglobin (Hb) has a slower rise at low pressures because the binding energy of a single O_2 to a molecule of Hb is lower than for Mb. At higher pressures of oxygen the Hb curve has a region that is concave upwards because the binding energy per O_2 increases after the first O_2 is adsorbed.

The O_2 molecules on hemes are in equilibrium with the O_2 in the surrounding liquid, so that the chemical potentials of O_2 are equal on the myoglobin and in solution:

$$\mu(MbO_2) = \mu(O_2), \lambda(MbO_2) = \lambda(O_2) \tag{5.67}$$

where $\lambda \equiv \exp(-\varepsilon/\tau)$. From Chapter 3 we lind the value of λ in terms of the gas pressure by the relation

$$\lambda = n/n_Q = p/\tau n_Q \tag{5.68}$$

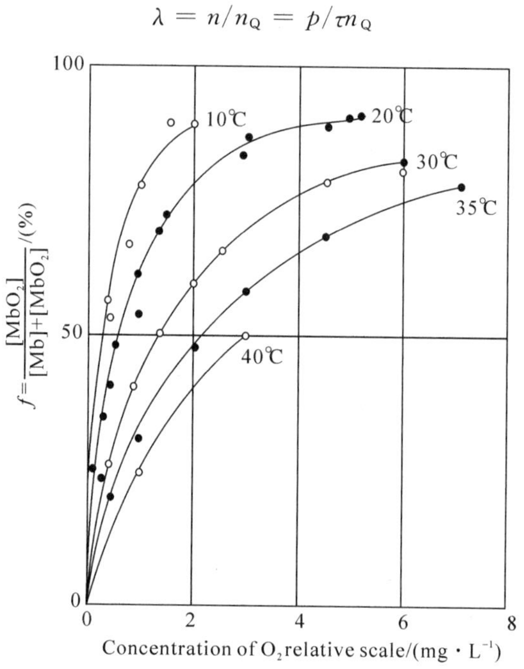

Figure 5.11 The reaction of a myoglobin (Mb) molecule with oxygen may be viewed as the adsorption of a molecule of O_2 at a site on the large myoglobin molecule

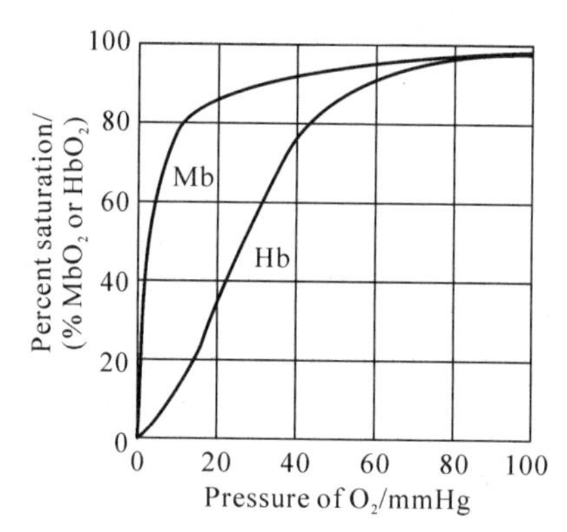

Figure 5.12 Saturation curves of O_2 bound to myoglobin (Mb) and hemoglobin (Hb) molecules in solution in water

We assume the ideal gas result appliesto O_2 in solution. At constant temperature $\lambda(O_2)$ is directly proportional to the pressure p.

The action f of Mb occupied by O_2 is found from Eq. (5.66) to be

$$f = \frac{\lambda \exp(-\varepsilon/\tau)}{1+\lambda \exp(-\varepsilon/\tau)} = \frac{1}{\lambda^{-1}\exp(-\varepsilon/\tau)+1} \tag{5.69}$$

We substitute Eq. (5.68) in Eq. (5.69) to obtain

$$f = \frac{1}{(n_Q\tau/p)\exp(\varepsilon/\tau)+1} = \frac{p}{n_Q\tau\exp(\varepsilon/\tau)+p} \tag{5.70}$$

or, with $p_0 \equiv n_Q\tau\exp(\varepsilon/\tau)$,

$$f = \frac{p}{p_0+p} \tag{5.71}$$

where p_0 is constant with respect to pressure but depends on the temperature. The result of Eq. (5.71) is known as the Langmuir adsorption isotherm when used to describe the adsorption of gases on the surfaces of solids.

Example: Impurity atom ionization in a semiconductor

Atoms of numerous chemical elements when present as impurities in a semiconductor may lose an electron by ionization to the conduction band of the semiconductor crystal. In the conduction band the electron moves about much as if it were a free particle and the electron gas in the conduction band may often be treated as an ideal gas. The impurity atoms are small systems s in thermal and diffusive equilibrium with the large reservoir formed by the rest of the semiconductor; the atoms exchange electrons and energy with the semiconductor.

Let I be the ionization energy of the impurity atom. We suppose that one but only one electron can be bound to an impurity atom: either orientation ↑ or ↓ of the electron spin is accessible. Therefore the system s has three allowed states one without an electron one with an electron attached with spin ↑, and one with an electron attached with spin ↓. When s has zero electrons the impurity atom is ionized. We choose the zero of energy of s as this state; the other two states therefore have the common energy $\varepsilon = -I$. The accessible states of s are summarized below(see Table 5.2).

Table 5.2 The accessible state of *s*

State number	Description	N	ε
1	Electron detached	0	0
2	Electron attached, spin ↑	1	$-I$
3	Electron attached, spin ↓	1	$-I$

The Gibbs sum is given by

$$\mathfrak{z} = 1 + 2\exp[(\mu+I)/\tau] \tag{5.72}$$

The probability that s is ionized ($N=0$) is

$$P(\text{ionized}) = P(0,0) = \frac{1}{\mathfrak{z}} = \frac{1}{1+2\exp[(\mu+I)/\tau]} \tag{5.73}$$

The probability that s is neutral (un-ionized) is

$$P(\text{neutral}) = P(1\uparrow, -I) + P(1\uparrow, -I) \tag{5.74}$$

which is just $1-P(0, 0)$.

5.5 Summary

(1) The chemical potential is defined as $\mu(\tau, V, N) \equiv (\partial F/\partial N)_{\tau,N}V$ and may also be found from $\mu = (\partial U/\partial N)_{\sigma,V} = \tau(\partial\sigma/\partial N)_{U,V}$. Two systems are in diffusive equilibrium if $\mu_1 = \mu_2$.

(2) The chemical potential is made up of two parts external and internal. The external part is the potential energy of a particle in an external field of force. The internal part is of thermal origin; for an ideal monatomic gas $\mu(\text{int}) = \tau\ln(n/n_Q)$ where n is the concentration and $n_Q \equiv (M\tau/2\pi h^2)^{3/2}$ is the quantum concentration.

(3) The Gibbs factor

$$P(N, \varepsilon_s) = \exp[(N\mu - \varepsilon_s)/\tau]/\mathfrak{z}$$

gives the probability that a system at chemical potential μ and temperature τ will have N particles and be in a quantum state s of energy ε_s.

(4) The Gibbs sum

$$\mathfrak{z} \equiv \sum_{\text{ASN}} \exp[(N\mu - \varepsilon_{s(N)})/\tau]$$

is taken over all states for all numbers of particles.

(5) The absolute activity λ is defined by $\lambda \equiv \exp(\mu/\tau)$.

(6) The thermal average number of particles is

$$\langle N \rangle = \lambda \frac{\partial}{\partial\lambda}\ln\mathfrak{z}$$

5.6 Problems

1. Centrifuge

A circular cylinder of radius R rotates about the long axis with angular velocity. The cylinder contains an ideal gas of atoms of mass M at temperature τ. Find an expression for the dependence of the concentration $n(r)$ on the radial distance r from the axis in terms of $n(0)$ on the axis. Take μ as for an ideal gas.

2. Molecules in the Earth's atmosphere

If n is the concentration of molecules at the surface of the Earth M the mass of a molecule and g the gravitational acceleration at the surface, show that at constant temperature the total number of molecules in the atmosphere is

$$N = 4\pi n(R)\exp(-MgR/\tau)\int_R^g \mathrm{d}r r^2 \exp(MgR^2/r\tau) \qquad (5.75)$$

with r measured fromthe center of the Earth: here R is the radius of the Earth. The integral diverges at the upper limit, so that N cannot be bounded and the atmosphere cannot be in equilibrium. Molecules, particularly light molecules are always escaping from the atmosphere.

3. Potential energy of gas in gravitational field

Consider a column of atoms each of mass M at temperature τ in a uniform gravitational field g. Find the thermal average potential energy per atom. The thermal average kinetic energy density is independent of height. Find the total heat capacity per atom. The total heat capacity is the sum of contributions from the kinetic energy and from the potential energy. Take the zero of the gravitational energy at the bottom $h=0$ of the column. Integrate from $h=0$ to $h=\infty$.

4. Active transport

The concentration of potassium K^+ ions in the internal sap of a plant cell (for example a fresh water alga) may exceed by a factor of 10^4 the concentration of K^+ ions in the pond water in which the cell is growing. The chemical potential of the K^- ions is higher in the sap because their concentration n is higher there. Estimate the difference in chemical potential at 300 K and show that it is equivalent to a voltage of 0.24 V across the cell wall.

Take μ as for an ideal gas. Because the values of the chemical potentials are different the ions in the cell and in the pond are not in diffusive equilibrium. The plant cell membrane is highly impermeable to the passive leakage of ions through it. Important questions in cell physics include these: How is the high concentration of ions built up within the cell? How is metabolic energy applied to energize the active ion transport?

5. Magnetic concentration

Determine the ratio m/τ for which Figure 5.6 is drawn. If $T=300$ K how many Bohr magnetons $\mu_H = e\hbar/2me$ would the particles contain to give a magnetic concentration effect of the magnitude shown?

6. Gibbs sum for a two level system

(1)Consider a system that may be un-occupied with energy zero or occupied by one particle in either of two states one of energy zero and one of energy ε. Show that the Gibbs sum for this system is

$$\mathfrak{z} = 1 + \lambda + \lambda \exp(-\varepsilon/\tau) \tag{5.76}$$

Our assumption excludes the possibility of one particle in each state at the same time. Notice that we include in the sum a term for $N=0$ as a particular state of a system of a variable number of particles.

(2) Show that the thermal average occupancy of the system is

$$\langle N \rangle = \frac{\lambda + \lambda \exp(-\varepsilon/\tau)}{\mathfrak{z}} \tag{5.77}$$

(3) Show that the thermal average occupancy of the state at energy ε is

$$\langle N(\varepsilon) \rangle = \lambda \exp(-\varepsilon/\tau)/\mathfrak{z} \tag{5.78}$$

(4) Find an expression for the thermal average energy of the system.

(5) Allow the possibility that the orbital at 0 and at ε may be occupied each by one particle at the same time; show that

$$\mathfrak{z} = 1 + \lambda + \lambda \exp(-\varepsilon/\tau) + \lambda^2 \exp(-\varepsilon/\tau) = (1+\lambda)[1 + \lambda \exp(-\varepsilon/\tau)] \tag{5.79}$$

Because $\mathfrak{z}$ can be factored as shown, we have in effect two independent systems.

7. States of positive and negative ionization

Consider a lattice offixed hydrogen atoms; suppose that each atom can exist in four states (see Table 5.3):

Table 5.3 Four states of each atom

State	Number of electrons	Energy
Ground	1	$-\frac{1}{2}\Delta$
Positive ion	0	$-\frac{1}{2}\delta$
Negative ion	2	$\frac{1}{2}\delta$
Excited	1	$\frac{1}{2}\Delta$

Find the condition that the average number of electrons per atom be unity. The condition will involve δ, λ, and τ.

8. Carbon monoxide poisoning

In carbon monoxide poisoning the CO replaces the O_2 adsorbed on hemoglobin (Hb) molecules in the blood. To show the effect, consider a model for which each adsorption site on a heme may be vacant or may be occupied either with energy ε_A by one molecule O_2 or with energy by one molecule CO. Let N fixed heme sites be in equilibrium with O_2 and CO in the gas phases at concentrations such that the activities are $\lambda(O_2)=1\times10^{-5}$ and $\lambda(CO)=1\times10^{-7}$ all atbody temperature 37℃. Neglect any spin multiplicity factors.

(1)First consider the system in the absence of CO. Evaluate CO such that 90 percent of the Hb sites are occupied by O_2. Express the answer in eV per O_2.

(2)Now admit the CO under the specified conditions. Find ε_B such that only 10 percent of the Hb sites are occupied by O_2.

9. Adsorption of O_2 in a magnetic field

Suppose that at most one O_2 can be bound to a heme group (see Problem 8), and that when $\lambda(O_2)=10^{-5}$ we have 90% of the hemes occupied by O_2. Consider O_2 as having a spin of 1 and a magnetic moment of 1 μ_B. How strong a magnetic field is needed to change the adsorption by 1 percent at $T=302$ K? (The Gibbs sum in the limit of zero magnetic field will differ from that of Problem 8 because there the spin multiplicity of the bound state was neglected.)

10. Concentration fluctuations

The number of particles is not constant in a system in diffusive contact with a reservoir. We have seen that

$$\langle N\rangle = \frac{\tau}{\mathfrak{z}}\left(\frac{\partial \mathfrak{z}}{\partial \mu}\right)_{\tau,V} \tag{5.80}$$

from Eq. (5.59).

(1)Show that

$$\langle N^2\rangle = \frac{\tau^2}{\mathfrak{z}}\left(\frac{\partial^2 \mathfrak{z}}{\partial \mu^2}\right)_{\tau,V} \tag{5.81}$$

The mean-square deviation $\langle(\Delta N)^2\rangle$of N from $\langle N\rangle$ is defined by

$$\left.\begin{aligned}\langle(\Delta N)^2\rangle &= \langle(N-\langle N\rangle)^2\rangle = \langle N^2\rangle - 2\langle N\rangle\langle N\rangle + \langle N\rangle^2 = \langle N^2\rangle - \langle N\rangle^2 \\ \langle(\Delta N)^2\rangle &= \tau^2\left[\frac{1}{\mathfrak{z}}\frac{\partial^2 \mathfrak{z}}{\partial \mu^2} - \frac{1}{\mathfrak{z}^2}\left(\frac{\partial \mathfrak{z}}{\partial \mu}\right)^2\right]\end{aligned}\right\} \tag{5.82}$$

(2)Show that this may be written as

$$\langle(\Delta N)^2\rangle = \tau\partial\langle N\rangle/\partial\mu \tag{5.83}$$

In Chapter 6 we apply this result tothe ideal gas to find that

$$\frac{\langle(\Delta N)^2\rangle}{\langle N\rangle^2} = \frac{1}{\langle N\rangle} \tag{5.84}$$

is the mean square fractional fluctuation in the population of anideal gas in diffusive contact with a reservoir. If $\langle N\rangle$ is of the order of 10^{20} atoms, then the fractional fluctuation is exceedingly small. In such a system the number of particles is well defined even though it cannot be rigorously constant because diffusive contact is allowed with the reservoir. When $\langle N\rangle$ is low, this relation can be used in the experimental determination of the molecular weight of large molecules such as DNA of MW 10^8-10^{10}; see M. Weissman, H. Schindler and G. Feher, Proc. Nat. Acad. Sci. 73, 2776 (1976).

11. Equivalent definition of chemical potential

The chemical potential was defined by Eq. (5.5) as $(\partial F/\partial N)_{\tau,V}$. An equivalent expression listed in Table 5.1 is

$$\mu = (\partial F/\partial N)_{\tau,V} \tag{5.85}$$

Prove that this relation which was used by Gibbs to define μ, is equivalent to the definition Eq. (5.5) that we have adopted. It will be convenient to make use of the results of Eq. (5.31) and Eq. (5.35). Our reasons for treating Eq. (5.5) as the definition of μ and Eq. (5.85) as a mathematical consequence, are two-fold. In practice we need the chemical potential more often as a function of the temperature τ than as a function of the entropy σ. Operationally, a process in which a particle is added to a system while the temperature of the system is kept constant is a more natural process than one in which the entropy is kept constant: Adding a particle to a system at a finite temperature tends to increase its entropy unless we can keep each system of the ensemble in a definite, although new quantum state. There is no natural laboratory process by which this can be done. Hence the definition Eq. (5.5) or Eq. (5.6), in which the chemical potential is expressed as the change in free energy per added particle under conditions of constant temperature is operationally the simpler. We point out that Eq. (5.85) will not give $U=\mu N$ on integration, because $\mu(N,\sigma,V)$ is a function of N.

12. Ascent of sap in trees

Find the maximum height to which water may rise in a tree under the assumption that the roots stand in a pool of water and the uppermost leaves are in air containing water vapor at a relative humidity $r=0.9$. The temperature is 25℃. If the relative humidity is r, the actual concentration of water vapor in the air at the uppermost leaves is rn_0 where n_0 is the concentration in the saturated air that stands immediately above the pool of water.

13. Isentropic expansion

(1) Show that the entropy of an ideal gas can be expressed as a function only of the orbital occupancies.

(2) From this result show that $\tau V^{2/3}$ is constant in an isentropic expansion of an ideal monatomic gas.

14. Multiple binding of O_2

A hemoglobin molecule can bind four O_2 molecules. assume that $\tau V^{2/3}$ is the energy of each bound O_2 relative to O_2 at rest at infinite distance. Let λ denote the absolute activity exp (μ/τ) of the free O_2 (in solution).

(1) What is the probability that one and only one O_2 is adsorbed on a hemoglobin molecule? Sketch the result qualitatively as a function of λ.

(2) What is the probability that four and only four O_2 are adsorbed? Sketch this result also.

15. External chemical potential

Consider a system at temperature τ with N atoms of mass M in volume V. Let $\mu(0)$ denote the value of the chemical potential at the surface of the earth.

(1) Prove carefully and honestly that the value of the total chemical potential for the identical system when translated to altitude h is

$$\mu(h) = \mu(0) + Mgh$$

where g is the acceleration of gravity.

(2) Why is this result different from that applicable to the barometric equation of an isothermal atmosphere?

Chapter 6 Ideal Gas

The ideal gas is a gas of noninteracting atoms in the limit of low concentration. The limit is defined below in terms of the thermal average value of the number of particles that occupy an orbital. The thermal average occupancy is called the distribution function, usually designated as $f(\varepsilon,\tau,u)$, where ε is the energy of the orbital.

An orbital is a state of the Schro dinger equation for only one particle. This term is widely used particularly by chemists. If the interactions between particles are weak, the orbital model allows us to exact quantum state of the Schro dinger equation of a system of N particles in terms of an approximate quantum state that we construct by assigning the N particles to orbitals, with each orbital a solution of a one-particle Schrodinger equation. There are usually an infinite number of orbitals available for occupancy. The term "orbital" is used even when there is no analogy to a classical orbit or to a Bohr orbit. The orbital model gives an exact solution of the N-particle problem only if there are no interactions between the particles.

It is a fundamental result of quantum mechanics (the derivation of which would lead us to astray here) that all species of particles fall into two distinct classes, fermions and bosons. Any particle with half-integral spin is a fermion, and any particle with zero or integral spin is a boson. There are no intermediate classes. Composite particles follow the same rule: an atom of ^{3}He is composed of an odd number of particles—2 electrons, 2 protons, 1 neutron each of spin $\frac{1}{2}$, so that ^{3}H must have half-integral spin and must be a fermion. An atom of ^{4}He has one more neutron, so there are an even number of particles of spin $\frac{1}{2}$, and ^{4}He must be a boson.

The fermion or boson nature of the particle species that make up a manybody system has a profound and important effect on the states of the system. The results of quantum theory as applied to the orbital model of noninteracting particles appear asoccupancy rules:

(1) An orbital can be occupied by any integral number of bosons of the same species, including zero.

(2) An orbital can be occupied by 0 or 1 fermion of the same species.

The second rule is a statement of the Pauli exclusion principle. Thermal averages of occupancies need not be integral or half-integral, but the orbital occupancics of any individual system must conform to one or the other rule.

The two different occupancy rules give rise to two different Gibbs sums for each orbital: there is a boson sum over all integral value of the orbital occupancy N, and there is a fermion sum in which $N=0$ or $N=1$ only. Different Gibbs sums lead to different quantum distribution functions $f(\varepsilon,\tau, u)$ for the thermal average occupancy . If conditions are such that $f<1$, it will not matter whether the occupancy $N=2,3,\cdots$ are excluded or are allowed. Thus when $f<1$ the fermion and boson distribution functions must be similar. This limit in which the orbital occupancy is small in comparison with unity is called the classical regime.

We now treat the Fermi-Dirac distribution functions for the thermal average occupancy of an orbital by fermions and the Bose-Einstein distribution function for the thermal average occupancy of an orbital by bosons. We show the equivalence of the two functions in the limit. We treat the properties of fermion and boson gases in the opposite limit, where the nature of the particles is absolutely crucial for the gas.

6.1 Fermi-Dirac Distribution Function

We consider a system composed of a single orbital that may be occupied by a fermion. The system is placed in thermal and diffusive contact with a reservoir, as in Figures 6.1 and 6.2 A real system may consist of a large number N_0 of fermions, but it is very helpful to focus on one orbital and call it the system. All other orbitals of the real system are thought of as the reservoir. Our problem is to find the thermal average occupancy of the orbital thus singled out. An orbital can be occupied by zero or by one fermion. No other occupancy is allowed by the zero if the orbital is unoccupied. The energy is ε if the orbital is occupied by one fermion.

The Gibbs sum now is simple: from the definition in Chapter 5 we have

$$\mathfrak{z} = 1+\lambda\exp(-\varepsilon) \tag{6.1}$$

The Eq. (6.1) comes from the configuration with occupancy $N=0$ and energy $\varepsilon=0$. The term $\lambda\exp(-\varepsilon)$ comes when the orbital is occupied by one fermion, so that $N=1$ and the energy is ε. The thermal averagc value of the occupancy of the orbital is the ratio of the term in the Gibbs sum with $N=$to the entire Gibbs sum:

$$\langle N(\varepsilon)\rangle = \frac{\lambda\ \exp(-\varepsilon/\tau)}{1+\exp(-\varepsilon/\tau)} = \frac{1}{\lambda^{-1}\ \exp(\varepsilon/\tau)+1} \tag{6.2}$$

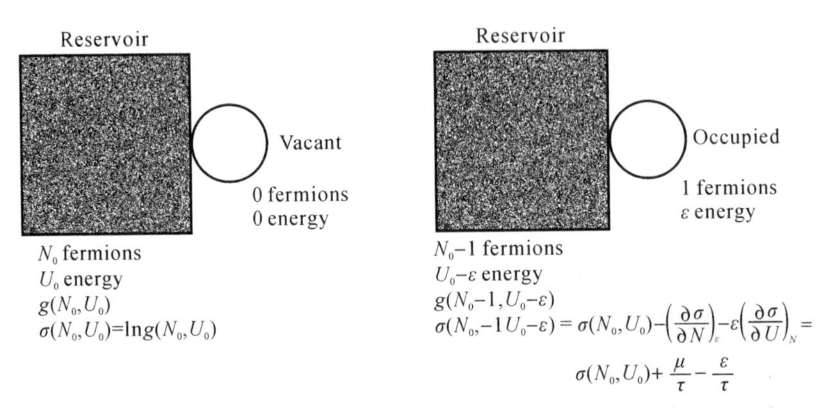

Figure 6. 1　We consider as the system a single orbital that may be occupied at most by one fermion

We introduce for the average occupancy the conventional symbol $f(\varepsilon)$ that denotes the thermal average number of particles in an orbital of energy ε:

$$f(\varepsilon) \equiv < N(\varepsilon) > \tag{6.3}$$

Recall from Chapter 5 that $\lambda \equiv \exp(\mu/\tau)$, where μ is the chemical potential. We may write Eq. (6. 2) in the standard form:

$$f(x) = \frac{1}{\exp[(\varepsilon - \mu)/\tau] + 1} \tag{6.4}$$

This result is known as the Fermi-Dirac distribution function. Eq. (6. 4) gives the average number of fermions in a single orbital of energy ε.

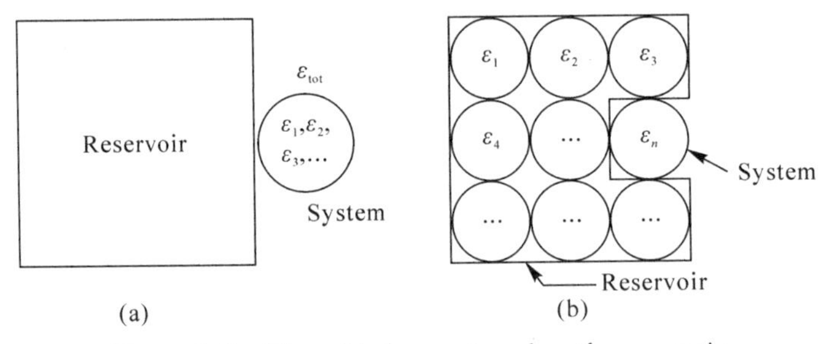

Figure 6. 2　The orbitals are viewed as the reservoir

(a) The obvious method of viewing a system of noninteracting particles is shown here; (b) A single orbital as the system

The energy levels each refer to an orbital that is a solution of a singleparticle Schrodinger equation. The total energy of the system is

$$\varepsilon_{tol} = \sum N_a \varepsilon_m$$

where N_a is the number of particles in the orbital n of energy ε_m. For fermions $N_a = 0$ or 1. Figure 6. 2(b) is much simpler than Figure 6. 2(a), and equally valid, to treat a single orbital as the system. The system in this scheme may be the orbital n of ε_m. All other orbitals are viewed as the reservior. The total energy of this one-orbital system is $N_m \varepsilon_m$, where N_m is the number of particles in the orbital. This device of using one orbital as the system works because the particles are supposed to interact only weakly with each other. If we think of the fermion system associated with the orbital n, these system has 1 particle and energy ε_m, Thus, the Gibbs sum consists of only two terms:

$$\mathfrak{z} = 1 + \lambda \exp(-\varepsilon_m)$$

The first term arises from the orbital occupancy $N_m = 0$, and the second term arises from $N_m = 1$. f always lies between zero and one. The Fermi-Dorac distribution function is plotted in Figure 6.3.

In the field of solid state physics the chemical potential μ is often called the Fermi level. The chemical potential usually depends on the temperature. The value of μ at zero temperature is often written as ε_F, that is

$$\mu \equiv \mu(0) = \varepsilon_F \tag{6.5}$$

We call ε_F the Fermi energy, not to be confused with the Fermi level which is the temperature dependent $\mu(\tau)$.

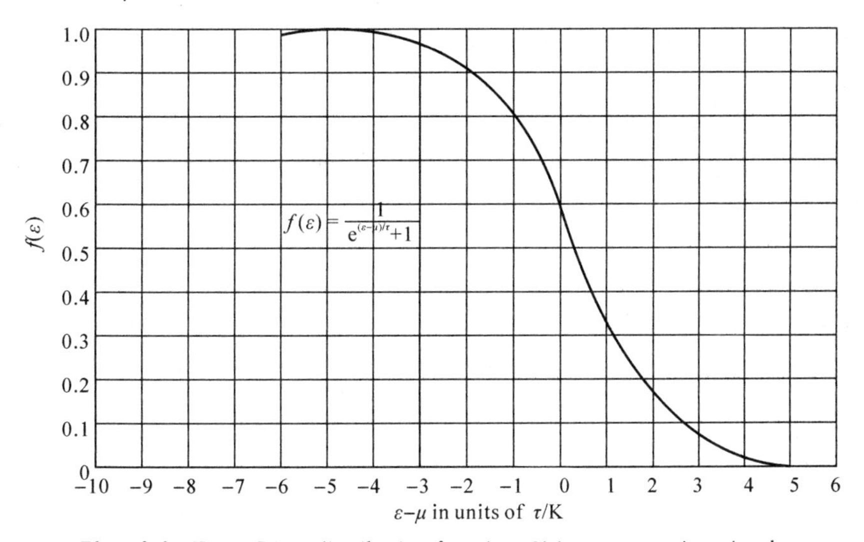

Figure 6.3 Plot of the Fermi-Dirac distribution function $f(\varepsilon)$ versus ε-μ in units the temperature

Consider a system of many independent orbitals, as in Figure 6.4. Al the temperature all orbitals of energy below the Fermi energy are occupied by exactly one fermion each, and all orbitals of higher energy are unoccupied. At nonzero temperatures the value of the chemical potential μ departs from the Fermi energy.

If there is an orbital of energy equal to the chemical potential ($\varepsilon = \mu$), the orbital is exactly half-filled, in the sense of a thermal average:

$$f(\varepsilon = \mu) = \frac{1}{1+1} \tag{6.6}$$

Orbitals of lower energy are more than half-filled, and orbitals of higher energy are less than half-filld.

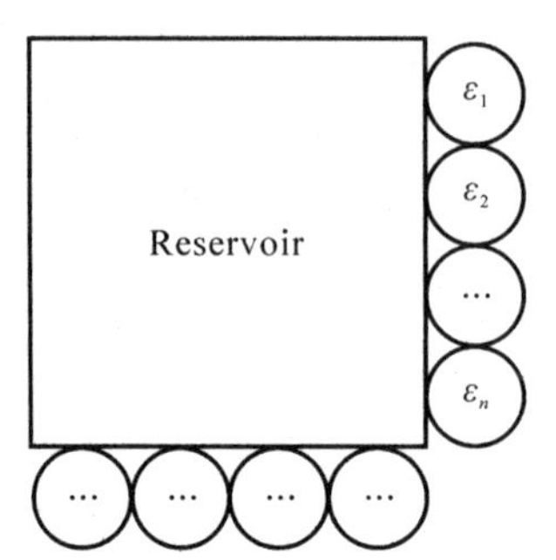

Figure 6.4 A convenient pictorial way to think of a system composed of independent orbitals that do not interact with each other, but interact with a common reservior

A boson is a particle with an integral value of spin. The occupancy rule for bosons is that an orbital can be occupied by any number of bosons, so that bosons have an essentially different quality than fermions Systems of bosons can have rather different physical properties than systems of fermions . Atoms of ^{4}He are bosons; atoms of ^{3}He are fermions. The remarkable superfluid properties of the low temperature ($T < 2.17$ K) phase of liquid helium can be attributed to the properties of a boson gas. There is a sudden increase in the fluidity and in the heat conductivity of liquid ^{4}He below this temperature. In experiments by Kapitza the flow viscosity of ^{4}He below the 2.17 K was found to be less than 10^{-7} of the viscosity of the liquid above 2.17 K.

Photons (the quata of the electromagnetic field) and phonons (the quanta of elastic waves in solids) can be considered to be bosons whose number is not conserves, but it is simpler to think of phonons as excitations of an oscillator, as we did in Chapter 4.

We consider the distribution function for a system of noninteracting bosons in thermal and diffusive contact with a reservoir. We assume the bosons are all of the same species. Let ε denote the energy of a single orbial when occupied by one particle; when there are N particles in the orbital, the energy is N_ε, as in Figure 6.5. We treat one orbial as the system and view all other orbitals as part of the reservoir.

Here ε is the energy of an orbital when occupied by one particle; N_ε is the energy of the same orbital when occupied by N particle. Any number of bosons can occupy the same orbital. The lowest level of this orbital contributes an Eq. (6.1) to the grand sum

$$\mathfrak{z} = \sum_{N=0}^{\infty} \lambda^N \exp(-N\varepsilon/\tau) = \sum_{N=0}^{\infty} [\lambda \exp(-\varepsilon/\tau)]^N \tag{6.7}$$

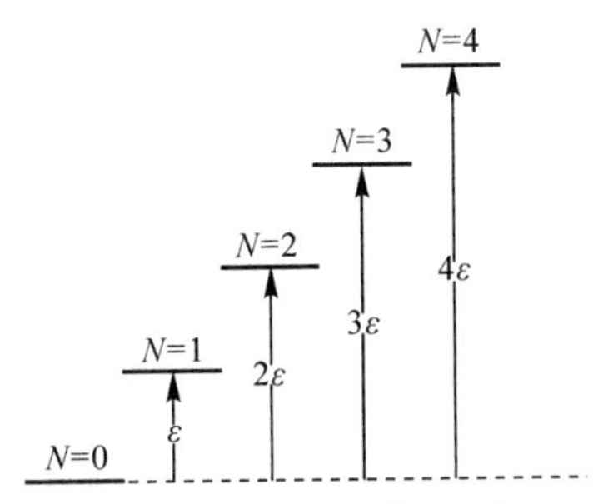

Figure 6. 5 Energy-level scheme fornoninteracting bosons

The upper limit on N should be the total number of particles in the combined system and reservior. However, the reservior may be arbitrarily large, so that N may run from zero to infinity. The Eq. (6. 7) may be summed in closed form. Let $x \equiv \lambda \exp(-\varepsilon/\tau)$; then

$$\mathfrak{z} = \sum_{N=0}^{\infty} x^N = \frac{1}{1-x} = \frac{1}{1-\lambda\ \exp(-\varepsilon/\tau)} \tag{6. 8}$$

provided that $\lambda \exp(-\varepsilon) < 1$. In all applications, $\lambda \exp(-\varepsilon)$ will satisfy this inequality; otherwise the number of bosons in the system would not be bounded.

The thermal average of the number of particles in the orbital is found from the Gibbs sum by use of Eq. (5. 62):

$$f(\varepsilon) = \lambda \frac{\partial}{\partial \lambda} \ln \mathfrak{z} = -x \frac{\mathrm{d}}{\mathrm{d}x} \ln(1-x) = \frac{x}{1-x} = \frac{1}{\lambda^{-1} \exp(\varepsilon/\tau) - 1} \tag{6. 9}$$

or

$$f(\varepsilon) = \frac{1}{\exp[(\varepsilon - u)/\tau] - 1} \tag{6. 10}$$

The classical regime is attained for $(\varepsilon - \mu) \gg 0$, where the two distributions become nearly identical.

This defines the Bose-Einstein distribution function. It differs mathematically from the Fermi-Dirac distribution function only by having -1 instead of $+1$ in the denominator . The charge can have very significant physical consequences. The two distribution functions are compared in Figure 6. 6. The ideal gas represents the limit $\varepsilon - \mu \gg 0$ in which the two distribution functions are approximately equal, as discussed below. The particular choice made in any problem will affect the value of the chemical potential μ, but the value of the difference $\varepsilon - \mu$ has to independent of the choice of the zero of ε, this point is discussed further in Eq. (6. 20) below.

A gas is in the classical regime when the average number of atoms in each orbital is much less than one. The average orbital occupancy for a gas at room temperature and atmospheric pressure is of the order of only 10^{-6}, safely in the classical regime. Differences between fermions (half-integral spin) and bosons arise only for occupancies of the order of one or more, so that in the classical regime their equilibrium properties are identical, The quantum regime is the opposite of the classical regime. These characteristic features are summarized in Table 6. 1.

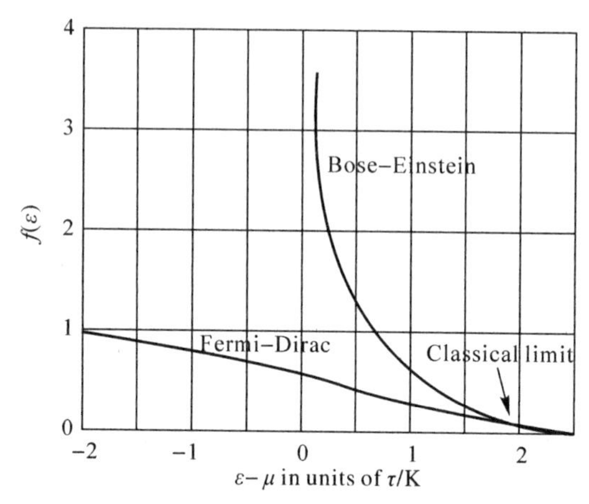

Figure 6. 6 Comparison of Bose-Einstein and Fermi-Dirac distribution functions

Table 6. 1 Comparison of the orbital occupancies in the classical and the quantum regimes

Regime	Class of particle	Thermal average occupancy of any orbital
Classical	Fermion	Always much less than one
	Boson	Always much less than one
Quantum	Fermion	Close to but less than one
	Boson	Orbital of lowest energy has an occupancy much greater than one

An ideal gas is defined as a system of free noninteraacting particles in the classical regime. "Free" means confined in box with no restrictions or external forces acting within the box. We develop the properties of an ideal gas with the use of the powerful method of the Gibbs sum. In Chapter 3 we treated the ideal gas by use of the partition function, but the identical particle problem encountered there was resolved by a method whose validity was not perfectly clear.

The Fermi-Dirac and Bose-Einstein distribution functions in the classical limit lead to the identical result for the average number of atoms in an orbital. Write $f(\varepsilon)$ for the average occupancy of an orbital at energy ε. Here ε is the energy of an orbital occupied by one particle; it is not the energy of a system of N particles. The Fermi-Dirac (FD) and Bose-Einstein (BE) distribution functions are

$$f(\varepsilon) = \frac{1}{\exp[(\varepsilon - \mu)/\tau] \pm 1} \tag{6.11}$$

where the plus sign is for the FD distribution and the minus sign for the BE distribution. In order that $f(\varepsilon)$ be much smaller than unity for all orbitals, we must have in this classical regime

$$\exp[(\varepsilon - \mu)/\tau] > 1 \tag{6.12}$$

for all ε. When this inequality is satisfied we may neglect the term ± 1 in the denominator of Eq. (6.11). Then for either fermions or bosons, the average occupancy of an orbital of energy

$$f(\varepsilon) = \exp[(\mu - \varepsilon)] = \lambda \exp(-\varepsilon) \tag{6.13}$$

with $\lambda = \exp(\mu)$. The limiting result of Eq. (6.13) is called the classical distribution function. It is the limit of the Fermi-Dirac and Bose-Einstein distribution functions when the average occupancy $f(\varepsilon)$ is very small in comparison with unity. Eq. (6.13), although called classical, is still a result for particles described by quantum mechanics: we shall find that the expression for λ or μ always involves the quantum contains $\hbar$. Any theory which contains $\hbar$ cannot be a classical theory.

We use the classical distribution function $f(\varepsilon) = \lambda \exp(-\varepsilon)$ to study the thermal properties of the ideal gas. There are many topics of importance: the entropy, chemical potential, heat capacity, the pressure-volume-temperature relation, and the distribution of atomic velocities. To obtain results from the classical distribution function, we need first to find the potential in term of the concentration of atoms.

The chemical potential is found from the condition that the thermal average of the total number of atoms equals the number of atoms known to be present. This number must be the sum over all orbitals of the distribution function $f(\varepsilon_s)$:

$$N = \langle N \rangle = \sum_s f(\varepsilon_s) \tag{6.14}$$

where s is the index of an orbital of energy ε_s. We start with a monatomic gas of N identical atoms of zero spin, and later we include spin and molecular modes of motion. The total number of atoms is the sum of the average number of atoms in each orbital. We use Eq. (6.13) in Eq. (6.14) to obtain

$$N = \lambda \sum_s \exp(-\varepsilon_s/\tau) \tag{6.15}$$

To evaluate this sum, observe that the summation over free particle orbitals is just the partition function Z_1 for a single free atom in volume V, whence $N = \lambda Z_1$.

In Chapter 3 it was shown that $Z_1 = n_Q V$, Where $n_Q \equiv (M\tau/2\pi\hbar^2)^{3/2}$ is the quantum concentration. Thus

$$N = \lambda Z_1 = \lambda n_Q V, \quad \lambda = N/n_Q V = n/n_Q \tag{6.16}$$

In terms of the number density $n = N/V$. Finally,

$$\lambda = \exp(\mu) = n/n_Q \tag{6.17}$$

Which is equal to the number of atoms in the quantum volume $1/n_Q$.

In the classical regime n/n_Q is $\ll 1$, The chemical potential of the ideal monatomic gas is

$$\mu = \ln(n/n_Q) \tag{6.18}$$

in agreement with Eq. (5.12a) obtained in anther way. The result may be written out to give

$$\mu = \tau[\ln N - \ln V - \frac{3}{2}\ln\tau + \frac{3}{2}\ln(2\pi\hbar/M)] \tag{6.19}$$

We see that the chemical potential increases as the concentration increases and decreases as the temperature increases.

Comment: The simple expression Eq. (6.18) for the chemical potential can be subject to several modification. We mention four examples.

If the zero of the energy scale is shifted by an energy Δ so that the zero of the kinetic energy of an orbital falls at $\varepsilon_0 = \Delta$ instead of at $\varepsilon_0 = 0$, then

$$\mu = \Delta + \ln(n/n_Q) \tag{6.20}$$

The chemical potential is related to the free energy by

$$(\partial F/\partial N)_{\tau,V} = \mu \tag{6.21}$$

From this,

$$F(N,\tau,V) = \int_0^N \mathrm{d}N\mu(N,\tau,V) = \tau\int_0^N \mathrm{d}N[\ln N + \cdots] \tag{6.22}$$

where the integrand is found in brackets in Eq. (6.19). Now

$$\int \mathrm{d}x \ln x = x\ln x - x$$

so that

$$F = N\tau[\ln N - \ln V - \frac{3}{2}\ln\tau + \frac{3}{2}\ln(2\pi\hbar^2/M)] \tag{6.23}$$

or

$$F = N\tau[\ln(n/n_Q) - 1] \tag{6.24}$$

The free energy increases with concentration and decreases with temperature.

Comment: The integral in Eq. (6.22) should strictly be a sum, because N a discrete variable. Thus, from Eq. (5.6)

$$F(N,\tau,V) = \sum_{N=1}^{N} \mu(N) \tag{6.25}$$

which differs from the integral only in the term in $\ln N$ in Eq. (6.19), for

$$\sum_{N=1}^{N} \ln N = \ln(1 \times 2 \times 3 \times \cdots \times N) = \ln N! \tag{6.26}$$

$$\ln N \approx N\ln N - N \tag{6.27}$$

may be use, and now Eq. (6.25) is the same as Eq. (6.23).

6.2 Pressure

The pressure is related to the free energy by Eq. (3.49):

$$p = -(\partial F/\partial N)_{\tau,n} \tag{6.28}$$

With Eq. (6.23) for F we have

$$p = N_\tau/V, \ pV = N_\tau \tag{6.29}$$

which is the idea gas law, as derived in Chapter 3.

The thermal energy U is found from $F \equiv U - \tau\sigma$, or

$$U = F + \tau\sigma = F - \tau(\partial F/\partial \tau)_{V,N} = -\tau^2\left(\frac{\partial F}{\partial \tau}\frac{F}{\tau}\right)_{V,N} \tag{6.30}$$

With Eq. (6.23) for F we have

$$\left(\frac{\partial}{\partial \tau}\frac{F}{\tau}\right)_{V,N} = \frac{-3N}{2\tau} \tag{6.31}$$

so that for an ideal gas

$$U = \frac{3}{2}N_\tau \tag{6.32}$$

The factor $\frac{3}{2}$ arises from the exponent of in n_Q because the gas is in three dimensions; if n_Q were in one or two dimensions, the factor would be $\frac{1}{2}$ or 1, respectively. The average kinetic energy of translation motion in the classical limit is equal to $\frac{1}{2}\tau$ or $\frac{1}{2}k_B T$ per translation degree of freedom of an atom. The principle of equipartition of energy among degrees of freedom was discussed in Chapter 3.

A polyatomic molecule has rotational degrees of freedom, and the average energy of each rotational degree of freedom is $\frac{1}{2}\tau$ when the temperature is high in comparison with the energy differences between the rotational energy levels of the molecule. The rotational energy is kinetic. A linear molecule has two degrees of rotational freedom which can be excited; a nonlinear molecule has three degrees of rotational freedom.

The entropy is related to the free energy by

$$\sigma = -(\partial F/\partial \tau)_{V,N} \tag{6.33}$$

From Eq. (6.23) for F we have the entropy of an ideal gas:

$$\sigma = N\left[\ln(n_Q/n) + \frac{5}{2}\right] \quad (6.34)$$

This is identical with our earlier result of Eq. (3.76). In the classical regime n/n_Q is$\ll 1$, so that $\ln(n/n_Q)$ is positive. The result of Eq. (6.34) is known as the Sackur-Tetrode equation for the absolute entropy of a monatomic ideal gas. It is important historically and is essential in the thermodynamics of chemical reactions. Even though the equation contains $\hbar$, the result was inferred from experiments on vapor pressure and on equilibrium in chemical reactions long before the quantum-mechanical basis was fully understood. It was a great challenge to theoretical physicists to explain the Sackur-Tetrode equation, and many unsuccessful attempts to do so were made in the early years of this century. We shall encounter application of the result in later chapters.

The entropy of the idea gas is directly proportional to the number of particles N if their concentration n is constant, as we see from Eq. (6.34). When two identical gases at identical conditions are placed side by side, each system having entropy σ_1, the total entropy is $2\sigma_2$ because N is doubled. If a value that connects the system is opened, the entropy is uncharged, we see that the entropy scales as the size of system: the entropy is linear in the number of particles at constant concentration. If the gases are not identical, the entropy increases when the value is opened (Problem 6).

The heat capacity at constant volume is defined in Chapter 3 as

$$C_V \equiv \tau(\partial\sigma/\partial\tau)_V \quad (6.35)$$

We can calculate the derivative directly from the entropy Eq. (6.34) of an ideal gas when the expression for n_Q is written out:

$$\left(\frac{\partial\sigma}{\partial\tau}\right) = \frac{\partial}{\partial\tau}\left(\frac{3}{2}N\ln\tau + \cdots\right) = \frac{3N}{2\tau}$$

From this, for an ideal gas

$$C_V = \frac{3}{2}N \quad (6.36)$$

or $C_V = \frac{3}{2}Nk_B$ in conventional units.

The heat capacity at constant pressure is large than C_V because additional heat must be added to perform the work needed to expend the volume of the gas against the constant pressure p. We use the thermodynamic identity $\tau d\sigma = dU + pdV$ to obtain

$$C_p = \tau\left(\frac{\partial\sigma}{\partial\tau}\right)_p = \left(\frac{\partial U}{\partial\tau}\right)_p + p\left(\frac{\partial V}{\partial\tau}\right)_p \quad (6.37)$$

The energy of an ideal gas depend only on the temperature, so that $(\partial U/\partial\tau)_p$ will have the same value as $(\partial U/\partial\tau)_V$, which is just C_V by the argument of Eq. (3.17b). By the ideal gas law $V=N_\tau/p$, so that the term $p(\partial U/\partial\tau)_p=N$, Thus Eq. (6.37) becomes

$$C_p = C_V + N \tag{6.38a}$$

In fundamental units, or

$$C_p = C_V + Nk_B \tag{6.38b}$$

In conventional units, we notice again the different dimensions that heat capacities have in the two system of units. For one mole, Nk_B is usually written as R, called the gas constant.

The result of Eq. (6.38a) and Eq. (6.38b) are written for an ideal gas without spin or other internal degrees of freedom of amolecule. For an atom $C_V=\frac{3}{2}N$, so that

$$C_p = \frac{3}{2}N + N = \frac{5}{2}N \tag{6.38c}$$

$$C_p = \frac{5}{2}Nk_B \tag{6.38d}$$

in conventional units, The ratio C_p/C_V is written as γ, the Greek letter gamma Example: Experimental tests of the Sackur-Tetrod equation, Experimental value of the entropy are often found from experimental values of C_p by numerical integration of Eq. (6.37) to give at constant pressure:

$$\sigma(\tau) - \sigma(0) = \int_0^\tau (C_p/\tau)\mathrm{d}\tau \tag{6.39}$$

The heat input required to melt the solid at 24.55 K is observed to be 355 J · mol^{-1}. The associated entropy of melting is

$$\Delta S_{\text{melting}} = \frac{335 \text{ J} \cdot \text{mol}^{-1}}{24.55 \text{ K}} = 13.64 \text{ J} \cdot \text{mol}^{-1} \cdot \text{K}^{-1}$$

The heat capacity of the liquid was measured from the melting point to the boiling point of 27.2 K under one atmosphere of pressure. The entropy increase was found to be

$$\Delta S_{\text{liquid}} = 3.85 \text{ J} \cdot \text{mol}^{-1} \cdot \text{K}^{-1}$$

The heat input required to vaporize the liquid at 27.2 K was observed to be 1,761 J · mol^{-1}. The associated entropy of vaporization is

$$\Delta S_{\text{vaporization}} = \frac{1{,}761 \text{ J} \cdot \text{mol}^{-1}}{27.2 \text{ K}} = 64.62 \text{ J} \cdot \text{mol}^{-1} \cdot \text{K}^{-1}$$

The experimental value of the entropy of neon gas at 27.2 K at a pressure of one atmosphere adds up to

$$S_{\text{gas}} = S_{\text{solid}} + \Delta S_{\text{melting}} + \Delta S_{\text{liquld}} + \Delta S_{\text{vaporization}} = 96.40 \text{ J} \cdot \text{mol}^{-1} \cdot \text{K}^{-1}$$

The calculated value of the entropy of neon under the same conditions is

$$S_{\text{gas}} = 96.45 \text{ J} \cdot \text{mol}^{-1} \cdot \text{K}^{-1}$$

from the Sackur-Tetrode equation. The excellent agreement with the experimental value gives us confidence in the basis of the entire theoretical apparatus us that led to Sackur-Tetrode equation. The result of Eq. (6.34) could hardly have been guessed; to find itverified by observation is a real experience. Results for argon and krypton are given in Table 6.2.

Table 6.2 Comparison of experiment and calculated values of the entropy at the boiling point under one atmosphere

Gas	T_{bp}/K	Entropy/($J \cdot mol^{-1} \cdot K^{-1}$)	
		Experimental	Calculated
Ne	27.2	96.40	96.25
Ar	87.29	129.75	129.24
Kr	119.93	144.56	145.06

We consider now an ideal gas of identical polyatomic molecules. Each molecule has rotation and vibrational degrees of freedom in addition to the translational degrees of freedom. The total energy ε of the molecule is the sum of two independent parts:

$$\varepsilon = \varepsilon_n + \varepsilon_{int} \tag{6.40}$$

where ε_{int} refers to the rotational and vibrational degrees of freedom and ε_n to the translational motion of the center of mass of the molecule. The vibrational energy problem is the harmonic oscillator problem treated earlier.

In the classical regime the Gibbs sum for the orbital n is

$$\xi = 1 + \lambda \exp(-\varepsilon_n/\tau) \tag{6.41}$$

where terms in higher powers of λ are omitted because the average occupancy of the orbital n is assumed to be $\ll 1$. That is, we neglect the terms in ξ which correspond to occupancies greater than unity. In the presence of internal energy states the Gibbs sum associated with the orbital n becomes

$$\xi = 1 + \lambda \sum_{int} \exp[-(\varepsilon_n + \varepsilon_{int})/\tau] \tag{6.42}$$

or

$$\xi = 1 + \lambda \exp(-\varepsilon_n/\tau) \sum_{int} \exp(-\varepsilon_{int}/\tau) \tag{6.43}$$

The summation is just the partition function of internal states:

$$Z_{int} = \sum_{int} \exp(-\varepsilon_{int}/\tau) \tag{6.44}$$

Which is related to the internal free energy of the one molecule by $F_{int} = -\tau \ln Z_{int}$.

From Eq. (6.43) the Gibbs sum is

$$\xi = 1 + Z_{int} \exp(-\varepsilon_n/\tau) \tag{6.45}$$

The probability that the translational orbital n is occupied, irrespective of the state of internal motion of the molecule, is given by the ratio of the term in λ to the Gibbs sum ξ:

$$f(\varepsilon_n) = \frac{\lambda Z_{int} \exp(-\varepsilon_n/\tau)}{1+\lambda Z_{int} \exp(-\varepsilon_n/\tau)} \approx \lambda Z_{int} \exp(-\varepsilon_n/\tau) \tag{6.46}$$

The classical regime was defined earlier as $f(\varepsilon_n) \ll 1$. The result of Eq. (6.46) is entirely analogous to Eq. (6.13) for the monatomic case, but λZ_{int} now plays the role of λ.

Several of the monatomic ideal gas are different for the polyatomic ideal gas:

(1) Eq. (6.17) for λ is replaced by

$$\lambda = n/(n_Q Z_{int}) \tag{6.47}$$

with n_Q defined exactly before (We shall always use n_Q as defined for the monatomic ideal gas of atoms with zero spin.). Because $\lambda \equiv \exp(\mu/\tau)$ we have

$$\mu = \tau[\ln(n/n_Q) - \ln Z_{int}] \tag{6.48}$$

(2) The free energy is increased by, for N molecules,

$$F_{int} = -N\tau \ln Z_{int} \tag{6.49}$$

(3) The entropy is creased by

$$\sigma_{int} = -(\partial F_{int}/\partial \tau)_V \tag{6.50}$$

The former result $U = \frac{3}{2} N\tau$ applies to the translational energy alone.

With the spin alone is

$$Z_{int} = (2I+1) \tag{6.51}$$

this being the number of independent spin states. The spin contribution to the free energy is

$$F_{int} = -\tau \ln(2I+1) \tag{6.52}$$

and the spin entropy is

$$\sigma_{int} = \ln(2I+1) \tag{6.53}$$

by Eq. (6.50). The effect of the spin entropy on the chemical potential is found with the help of Eq. (6.48):

$$\mu = \tau[\ln(n/n_Q) - \ln(2I+1)] \tag{6.54}$$

6.3 Reversible Isothermal Expansion

Consider as a model example 1×10^{22} atoms of ^4He at an initial volume of 10^3 cm^3 at 300 K. let the gas expand slowly at constant temperature until the volume is 2×10^3 cm^3. The temperature is maintained constant by thermal contact with a large reservoir. In a reversible expansion the system at any instant is in its most problem configuration.

1. What is the pressure after expansion?

The final volume is twice the initial volume; the final temperature is equal to the initial temperature. From $pV=N\tau$ we see that the final pressure is one-half the initial pressure.

2. What is the increase of entropy on expansion?

The entropy of an ideal gas at constant temperature depends on volume as

$$\sigma(V) = N\ln V + \text{constant} \tag{6.55}$$

Whence

$$\sigma_2 - \sigma_1 = N\ln(V_2 - V_1) = N\ln 2 = (1 \times 10^{22})(0.693) = 0.069 \times 10^{23} \tag{5.56}$$

When the gas expands isothermally, it done work against a piston, as in Figure 6.7. The work done on the piston when the volume is doubled is

$$\int_{V_1}^{V_2} p\mathrm{d}V = \int_{V_1}^{V_2} (N\tau/V)\mathrm{d}V = N\tau\ln(V_2/V_1) = N\tau\ln 2 \tag{6.57}$$

We evaluate $N\tau$ directly as 4.14×10^8 erg $=41.4$ J. Thus the work done on the piton is, form Eq. (6.57),

$$N\tau\ln 2 = (41.4\ \text{J})(0.693) = 28.7\ \text{J} \tag{6.58}$$

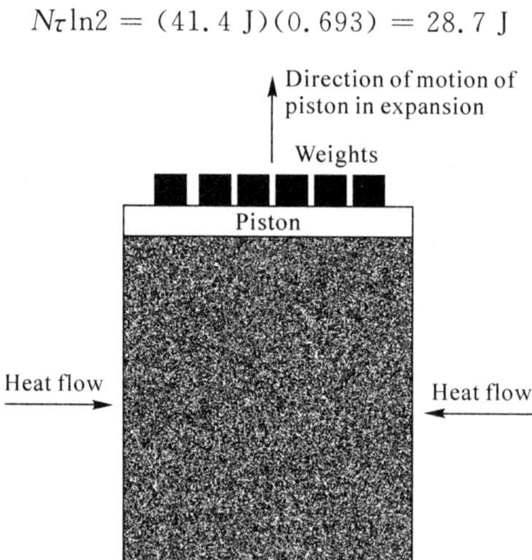

Figure 6.7 Work is done by the gas in an isothermal expansion

The assumption that the process is reversible enters in Eq. (6.57) when wen assume that a knowledge of V at every stage determines p at every stage of the expansion.

We define W as the work done on the gas by external agencies. This is the negative of the work done by the gas on the piston. From Eq. (6.58),

$$W = -\int p\mathrm{d}V = -28.7\ \text{J} \tag{6.59}$$

The energy of an ideal monatomic gas is $U=\frac{3}{2}N_\tau$ and does not change in an expansion at constant temperature. However, the Helmholtz free energy decreases.

3. How much heat flowed into the gas from the reservoir?

We have seen that the energy of the idea gas remained constant when the gas did work on the piston. By the conservation of energy it is necessary that a flow of energyin the form of heat into the gas occur from the reservoir through the walls of the container. The quantity Q of heat added to gas must be equal, but be opposite in sign, to the work done by the piston, because $Q+W=0$. Thus from the result of Eq. (6.59)

$$Q = 28.7\ \text{J} \tag{6.60}$$

6.4 Reversible Expansion at Constant Entropy

We considered above an expansion at constant temperature. Suppose instead that the gas expands reversibly from $1\times10^3\ \text{cm}^3$ to $2\times10^3\ \text{cm}^3$ in an insulated container. No heat flow to or from the gas is permitted, so that $Q=0$. The entropy is constant in a system isolated from the reservoir if the expansion process is carried out reversibly(slowly). A process without a change of entropy is called an isentropic process or an adiabatic process. The term "adiabatic" has the specific meaning that there is no heat transfer in the process. For simplicity, we shall stick with "isentropic".

1. What is the temperature of the gas after expansion?

The entropy of an ideal monatomic gas depends on the volume and the temperature as

$$\sigma(\tau,V) = N(\ln\tau^{\frac{3}{2}} + \ln V + \text{constant}) \tag{6.61}$$

So that the entropy remains constant if

$$\ln\tau^{\frac{3}{2}}V = \text{constant}, \quad \tau^{\frac{3}{2}}V = \text{constant} \tag{6.62}$$

In an expansion at constant entropy from V_1 to V_2 we have

$$\tau_1^{\frac{3}{2}}V_1 = \tau_2^{\frac{3}{2}}V_2 \tag{6.63}$$

for an ideal monatomic gas.

We use the ideal gas law $pV=N\tau$ to obtain two alternate forms. We insert $V=N\tau/p$ into Eq. (6.63) and cancel N on both sides to obtain

$$\frac{\tau_1^{\frac{5}{2}}}{p_1} = \frac{\tau_2^{\frac{5}{2}}}{p_2} \tag{6.64}$$

Similarly, we insert $\tau=pV/N$ in Eq. (6.63) to obtain

$$p_1^{\frac{3}{5}}V_1^{\frac{5}{3}} = p_2^{\frac{3}{5}}V_2^{\frac{5}{3}} \quad \text{or} \quad p_1 V_1^{\frac{5}{3}} = p_2 V_2^{\frac{5}{3}} \tag{6.65}$$

Both Eq. (6.64) and Eq. (6.65) hold only for a monatomic gas.

It is the subject of Problem 10 to generalize these results for an ideal gas of molecules with internal degrees of motion (rotations, vibrations). We obtain for an isentropic process

$$\tau_1 V_1^{\gamma-1} = \tau_2 V_2^{\gamma-1} \tag{6.66}$$

$$\tau_1^{\gamma/(1-\gamma)}\, p_1 = \tau_2^{\gamma/(1-\gamma)}\, p_2 \tag{6.67}$$

$$p_1 V_1^{\gamma} = p_2 V_2^{\gamma} \tag{6.68}$$

Here $\gamma \equiv \frac{C_p}{C_V}$ is the ratio of the heat capacities at constant pressure and constant volume.

With $T_1 = 300$ K and $V_1/V_2 = \frac{1}{2}$, we find from Eq. (6.63):

$$T_2 = \left(\frac{1}{2}\right)^{\frac{2}{3}} (300\ \text{K}) = 189\ \text{K} \tag{6.69}$$

This is the final temperature after the expansion at constant entropy. This gas is cooled in the expansion process by

$$T_1 - T_2 = 300\ \text{K} - 189\ \text{K} = 111\ \text{K} \tag{6.70}$$

Expansion at constant entropy is an important method of refrigeration.

1. What is the change in energy in the expansion?

The energy change is calculated from the temperature change of Eq. (6.70). For an ideal monatomic gas

$$U_2 - U_1 = C_V(\tau_2 - \tau_1) = \frac{3}{2}N(\tau_2 - \tau_1) \tag{6.71}$$

Sudden Expansion into a Vacuum or, in conventional units,

$$U_2 - U_1 = \frac{3}{2}Nk_B(T_2 - T_1) = \frac{3}{2}(1 \times 10^{22})(1.38 \times 10^{-16}\,\text{erg} \cdot \text{K}^{-1})(-111\ \text{K}) =$$
$$-2.3 \times 10^{8}\,\text{erg} = -23\ \text{J} \tag{6.72}$$

The energy decrease in an expansion at constant entropy. The work done by the gas is equal to the decrease in energy of the gas, which is $U_2 - U_1 = 23$ J.

6.5 Sudden Expansion into a Vacuum

Let the gas expand suddenly into a vacuum from an initial volume of 1 liter to a final volume of 2 liter. This is an excellent example of an irreversible process. When a hole is opened in the partition to permit the expansion, the first atoms rush through the hole and strike the opposite wall. If no heat flow through the walls is permitted, there is no way for the atoms to lose their kinetic energy. The subsequent flow may be turbulent(irreversible), with different parts of the gas at different values of the energy density. Irreversible energy flow between regions will eventually equalize conditions throughout the gas. We assume the whole process occurs rapidly enough so that no heat flows in though the walls.

1. How much work is done in the expansion?

No means of doing external work is provided, so that the work done is zero. Zero work is not necessarily a characteristic of all irreversible, but the work is zero for expansion a vacuum.

2. What is the temperature after expansion?

No work is done and no heat is added in the expansion: $W=0$, $Q=0$, and $U_2-U_1=0$. Because the energy is unchanged, the temperature of the ideal gas is unchanged. The energy of a real gas may change in the process because the atoms are moved farther apart, which affects their interaction energy.

3. What is the change of entropy in the expansion?

The increase of entropy when the volume is doubled at constant temperature is given by Eq. (6.56):

$$\Delta\sigma = \sigma_2 - \sigma_1 = N\ln 2 = 0.069 \times 10^{23} \tag{6.73}$$

For the expansion into a vacuum $Q=0$.

Expansion into a vacuum is not a reversible process: the system is not in the most probable(equilibrium) configuration at every stage of the expansion.

6.6 Problems

1. Derivative of Fermi-Dirac function

Show that $-\partial f/\partial\varepsilon$ evaluate at the Fermi level $\varepsilon=\mu$ has the value $(4\tau)^{-1}$. Thus the lower the temperature, the steeper the slope of the Fermi-Dirac function.

2. Symmetry of filled and vacant orbitals

Let $\varepsilon=\mu+\delta$, that $f(\varepsilon)$ appears as $f(\mu+\delta)$. Show that

$$f(\mu+\delta) = 1 - f(\mu-\delta) \tag{6.74}$$

Thus the probability that an orbital δ above the Fermi level is occupied is equal to the probability an orbital δ below the Fermi level is vacant. A vacant orbital is something know as a hole.

3. Distribute function for double occupancy statistics

Let us imagine a new mechanics in which the allowed occupancies of an orbital are 0, 1, and 2. The values of the energy associated with these occupancies are assumed to be 0, ε, and 2ε, respectively.

(1) Derive an expression for the ensemble average occupancy $<N>$, when the system composed of this orbital is in thermal an diffusive contact with a reservoir at temperature τ and chemical potential μ.

(2) Return now to the usual quantum mechanics, and derive an expression for the ensemble average occupancy of an energy level which is doubly degenerate; that is, two orbital have the identical energy ε. If both orbitals are occupied the total energy is 2ε.

4. Energy of gas of extreme relativistic particles

Extreme relativistic particles have momenta p such that $pc \gg Mc^2$, where M is the rest mass of the particle. The de Broglie relation $\lambda = h/p$ for the quantum wavelength continues to apply. Show that the mean energy per particle of an extreme relativistic ideal gas is 3τ if $\varepsilon \approx pc$, in contrast to $\frac{3}{2}\tau$ for the nonrelativistic problem (An interesting variety of relativistic problem are discussed by E. Fermi in Notes on Thermodynamics and Statistic, University of Chicago Press, 1996, paperback.).

5. Integration of the thermodynamic identity for an ideal gas

From the thermodynamic identity at constant number of particle we have

$$\mathrm{d}\sigma = \frac{\mathrm{d}U}{\tau} + \frac{p\mathrm{d}V}{\tau} = \frac{1}{\tau}\left(\frac{\partial U}{\partial \tau}\right)_V \mathrm{d}\tau + \frac{1}{\tau}\left(\frac{\partial U}{\partial V}\right)_\tau \mathrm{d}V + \frac{p\mathrm{d}V}{\tau} \tag{6.75}$$

Show by integration that for an ideal gas the entropy is

$$\sigma = C_V \ln\tau + N\ln V + \sigma_1 \tag{6.76}$$

where σ_1 is a constant, independent of τ and V.

6. Entropy of mixing

Suppose that a system of N atoms of type A is placed in diffusive contact with a system of N atoms of type B at the same temperature and volume. Show that after diffusive equilibrium is the total entropy is increase by $2N\ln 2$. The entropy increasing $2N\ln 2$ is know as the entropy of mixing. If the atoms are identical (A≡B), show that there is no increase in entropy when diffusive contact is established. The difference in the results has been called the Gibbs paradox.

7. Relation of pressure and energy density

(1)Show that the average pressure in a system in thermal contact with a heat reservoir is given by

$$p=-\frac{\sum_s(\partial\varepsilon_s/\partial V)_N\exp(-\varepsilon_s/\tau)}{Z} \tag{6.77}$$

where the sum is over all states of the system.

(2) Show for a gas of free particles that

$$\left(\frac{\partial\varepsilon_s}{\partial V}\right)_N=-\frac{2}{3}\frac{\varepsilon_s}{V} \tag{6.78}$$

as a result of the boundary condition of the problem. The result holds equally whether ε_s refers to a state of N noninteracting particles or to an orbital.

(3)Show that for a gas of free nonrelativistic particles

$$p=2U/3V \tag{6.79}$$

where U is the thermal average energy of the system. This result is not limited to the classical regime; it holds equally for fermion and boson particles, as long as they are nonrelativistic.

8. Time for a large fluctuation

We quoted Boltzmann to the effect that two gases in 0.1 L container will unmix only in a time enormously long compared to $10^{(10^{10})}$ years. We shall investigate a related problem: we let a gas of atoms of ^{4}He occupy a container of volume of volume of 0.1 L at 300 K and a pressure of 1 atm. and we ask how long it will be before the atoms assume a configuration in which all are in one-half of the container.

(1)Estimate the number of states accessible to the system in this initial condition.

(2)The gas is compressed isothermally to a volume of 0.05 L. How many states are accessible now?

(3)For the system in the 0.1 L container, estimate value of the ratio volume number of states for which all atoms are anywhere in the volume.

(4)If the collision rate of an atom is$\approx 10^{10}\ s^{-1}$, what is the total number of collisions of atoms in the system in a year? We use this as a crude estimate of the frequently with which the system changes.

(5) Estimate the number of years you would expect to wait before all atoms are in one-half of the volume, starting from the equilibrium configuration.

9. Gas of atoms with internal degree of freedom

Consider an ideal monatomic gas, but one for which the atom has two internal energy states, one an energy Δ above the other. There are N atoms in volume V at temperature τ. Find the ① chemical potential; ② free energy; ③ entropy; ④ pressure; ⑤ heat capacity at constant pressure.

10. Isentropic relations of ideal gas

(1) Show that the differential changes for an ideal gas in an isentropic process satisfy

$$\frac{\mathrm{d}p}{p}+\gamma\frac{\mathrm{d}V}{V}=0,\quad \frac{\mathrm{d}\tau}{\tau}+(\gamma-1)\frac{\mathrm{d}V}{V}=0,\quad \frac{\mathrm{d}p}{p}+\frac{\gamma}{1-\gamma}\frac{\mathrm{d}\tau}{\tau}=0 \tag{6.80}$$

where $\gamma=C_p/C_V$; these relations apply even if the molecules have internal degrees of freedom.

(2) The isentropic and isothermal bulk moduli are defined as

$$B_\sigma=-V(\partial p/\partial V)_\sigma,\quad B_\tau=-V(\partial p/\partial V)_\tau \tag{6.81}$$

Show that for an ideal gas $B_\sigma=\gamma p$; $B_\tau=p$. The velocity of sound in a gas is given by $c=(B_\sigma/\rho)^{1/2}$; there is very little heat transfer in a sound wave. For an ideal gas of molecules of massMwe have $p=\rho\tau/M$, so that $c=(\gamma\tau/M)^{1/2}$. Here ρ is the mass density.

11. Convective isentropic equilibrium of the atmosphere

The lower 10 - 15 km of the atmosphere—the troposphere—is often in a convective steady state at constant entropy, not constant temperature. In such equilibrium pV^γ is independent of altitude, where $\gamma=C_p/C_V$.

Use the condition of mechanical equilibrium in a uniform gravitational field to:

(1) Show that $\mathrm{d}T/\mathrm{d}z=$ constant, where z is the altitude. This quantity, important in meteorology, is called the dry adiabatic lapse rate (Do not use the barometric pressure relation that was derived in Chapter 5 for an isothermal atmosphere.).

(2) Estimate $\mathrm{d}T/\mathrm{d}z$, in C per km. Take $\gamma=7/5$.

(3) Show that $p\propto\rho^\gamma$, where ρ is the mass density.

If the actual temperature gradient is greater than the isentropic gradient, the atmosphere may be unstable with respect to convection.

12. Ideal gas in two dimensions

(1) Find the chemical potential of an ideal monatomic gas in two dimensions, with N at-

oms confined to a square of area $A=L^2$. The spin is zero.

(2) Find an expression for the energy U of the gas.

(3) Find an expression for the entropy σ. The temperature is τ.

13. Gibbs sum for ideal gas

(1) With the help of $Z_N=(n_Q V)^N/N!$ form Chapter 3, show that the Gibbs sum for an idea gas of identical atoms is $\xi=\exp(\lambda n_Q V)$.

(2) Show that the probability there are N atoms in the gas in volume V in diffusive contact with a reservoir is

$$P(N) = \langle N\rangle^N \exp(-\langle N\rangle)/N! \tag{6.82}$$

which is just the Poisson distribute function. Here $\langle N\rangle$ is the thermal average number of atoms in the volume, which we have evaluated previously as $\langle N\rangle=\lambda V n_Q$.

(3) Confirm that $P(N)$ above satisfies

$$\sum_N P(N) = 1 \quad \text{and} \quad \sum_N NP(N) = \langle N\rangle$$

14. Ideal gas calculations

Consider one mole of an ideal monatomic gas at 300 K and 1 atm. First, let this gas expand isothermally and reversibly to twice the initial volume; second, let this be followed be an isentropic expansion from twice to four times the initial volume.

(1) How much heat (in joules) is added to the gas in each of these two processes?

(2) What is the temperature at the end of the second process? Suppose the first process is replaced by an irreversible expansion into a vacuum, to a total volume twice the initial volume.

(3) What is the increase of entropy in the irreversible expansion. In joules per Kelvin?

15. Diesel engine compression

A diesel engine is an internal combustion engine in which fuel is sprayed into the cylinders after the air charge has been so highly compressed that it has attained a temperature sufficient to ignite the fuel. Assume that the air in the cylinders is compressed isentropically from an initial temperature of 27℃ (300 K). If the compression ratio 15, what is the maximum temperature in ℃ to which the air is heated by the compression? Take $\gamma=1.4$.

References

[1] UCHIDA K, NAKAMURA S, IRIE M. Thermally irreversible photochromic systems. Substituent effect on the absorption wavelength of 11,12-dicyano-5a, 5b-dihydro-5a,5b-dimethylbenzo [1,2-b, 6,5 - b']bis[1]benzothiophene[J]. B Chem Soc Jpn, 1992, 65: 430 - 435.

[2] IRIE M, MIYATAKE O, UCHIDA K, et al. Photochromic diarylethenes with intralocking arms[J]. J Am Chem Soc, 1994, 116: 9894 - 9900.

[3] IRIE M, SAKEMURA K, OKINAKA M, et al. Photochromism of dithienylethenes with electron- donating substituents[J]. J Org Chem, 1995, 60:8305 - 8309.

[4] MATSUDA K, IRIE M. Diarylethene as a photoswitching unit[J]. J Photoch Photobio C, 2004, 5:169 - 182.

[5] MATSUDA K, IRIE M. A diarylethene with two nitronyl nitroxides: Photoswitching of intramolecular magnetic interaction[J]. J Am Chem Soc, 2000, 122:7195 - 7201.

[6] KOSE M. Novel sulfoxide-introducing reaction and photochromic reactions of ethenylsulfinyl derivatives of dithienylethenes[J]. J Photoch Photobio A, 2004, 165:97 - 102.

[7] NAKATSUJI S. Recent progress toward the expoitation of organic radical compounds with photo-responsive magnetic properties[J]. Chem Soc Rev, 2004, 33:348 - 353.

[8] TIAN H, YANG S. Recent progress on diarylethene based photochromic switches[J]. Chem Soc Rev, 2004, 33:85 - 97.

[9] UCHIDA K, MATSUOKA T, KOBATAKE S, et al. Substituent effect on the photochromic reactivity of bis(2-thienyl) perfluorocyclopentenes[J]. Tetrahedron, 2001, 57: 4559 - 4565.

[10] LUCAS L N, ESCH J V, KELLOGG R M, et al. A new synthetic route to symmetrical photochromic diarylperfluorocyclopentenes[J]. Tetrahedron Lett, 1999, 40:1775 - 1778.

[11] CHEMLA D S, ZYSS J. Nonlinear optical properties of organic molecules and crystals [M]. Orlando: Academic Press, 1987.

[12] LEHN J M. Perspectives in supramolecular chemistry: from molecular recognition towards molecular information processing and self-organization[J]. Angew Chem Int Edit, 1990, 29:1304 - 1319.

[13] KANIS D R, RATNER M A, MARKS T J. Design and construction of molecular Assemblies with large second-order optical nonlinearities: Quantum chemical aspects[J]. Chem Rev, 1994, 94:195 - 242.
[14] GREEN M L H, MARDER S R, THOMPSON M E, et al. Synthesis and structure of (cis)-[1-ferrocenyl-2-(4 - nitrophenyl) ethylene], an organotransition metal compound with a large second-order optical nonlinearity[J]. Nature, 1987, 330:360 - 362.
[15] CHEN B, WANG M, WU Y, et al. Reversible near-infrared fluorescence switch by novel photochromic unsymmetrical-phthalocyanine hybrids based on bisthienylethene [J]. Chem Commun, 2002, 121: 1060 - 1061.
[16] GIORDANO L, JOVIN T M, IRIE M, et al. Diheteroarylethenes as thermally stable photoswitchable acceptors in photochromic fluorescence resonance energy transfer (pc-FRET)[J]. J Am Chem Soc, 2002, 124:7481 - 7489.
[17] MALY K E, WAND M D, LEMIEUX R P. Bistable ferroelectric liquid crystal photoswitch triggered by a dithienylethene dopant[J]. J Am Chem Soc, 2002, 124:7898 - 7899.
[18] TSUJIOKA T, KUME M, IRIE M. Optical density dependence of write/read characteristics in photon-mode photochromic memory[J]. Jpn J Appl Phys, 1996, 35:4353 - 4360.
[19] MYLES A J, BRANDA N R. 1,2-dithienylethene photochromes and non-destructibe erasable memory[J]. Adv Funct Mater, 2002, 12:167 - 173.
[20] CAMPBELL I H, DAVIDS P S, SMITH D L, et al. The schottky energy barrier dependence of charge injection in organic light-emitting diodes[J]. Appl Phys Lett, 1998, 72:1863 - 1865.
[21] SCHREIBER M, BUSS V. Origin of the bathochromic shift in the early photointermediates of the rhodopsin visual cycle: A CASSCF/CASPT2 study[J]. Int J Quantum Chem, 2003, 95:882 - 889.
[22] LIU Y J, HUANG M B. A theoretical study of low-lying singlet electronic states of SF2[J]. Chem Phys Lett, 2002, 360:400 - 405.
[23] GMEZ I, OLIVELLA S, REGUERO M, et al. Thermal and photochemical rearrangement of bicycle[3. 1. 0]hex-3-en-2-one to the ketonic tautomer of phenol: Computational evidence for the formation of a diradical rather than a zwitterionic intermediate[J]. J Am Chem Soc, 2002, 124:15375 - 15384.
[24] APLINCOURT P, HENON E, BOHR F, et al. Theoretical study of photochemical processes involving singlet excited states of formaldehyde carbonyl oxide in the atmosphere[J]. Chem Phys, 2002, 285:221 - 231.

[25] XENIDES D. On the performance of DFT methods in (hyper) polarizability calculations: N-4(T-d) as a test case[J]. J Mol Struc-Theochem, 2007, 804:41 - 46.

[26] AOTO Y A, ORNELLAS F R. Predicting new molecular species of potential interest to atmospheric chemistry: The isomers HSBr and HBrS[J]. J Phys Chem A, 2007, 111:521 - 525.

[27] SERRANO A L, MERCHAN M. Quantum chemistry of the excited state: 2005 overview[J]. J Mol Struc-Theochem, 2005, 729:99 - 108.

[28] WATTS J D, BARTLETT R J. Iterative and non-iterative triple excitation corrections in coupled-cluster methods for excited electronic states: the EOM-CCSDT-3 and EOM-CCSD(T) methods[J]. Chem Phys Lett, 1996, 258:581 - 588.

[29] SATTELMEYER K W, STANTON J F, OLSEN J, et al. A comparison of excited state properties for iterative approximate triple linear response coupled cluster methods [J]. Chem Phys Lett, 2001, 347:499 - 504.

[30] BRANDBYGE M, MOZOS J L, ORDEJON P, et al. Fluorescence switching of photochromic diarylethenes Phys[J]. Rev B, 2002, 65: 165401 - 165408.

[31] CRANO J C, GUGLIELMETTI R J. Organic photochromic and thermochromic compounds, Vol. 1: Main photochromic families[M]. New York: Plenum Press, 1999.

[32] MCCORMICK F B, ZHANG H, DVORNIKOV A, et al. Parallel access 3 - D multilayer optical storage using 2-photon recording[J]. SPIE, 1999, 3802:173 - 181.

[33] HUNTER S, KIAMILEV F, ESENER S, et al. Potentials of two-photon based 3 - D optical memories for high performance computing[J]. Appl Optics, 1990, 29:2058 - 2066.

[34] PARTHENOPOULOS D A, RENTZEPIS P M. Three-dimensional optical storage memory[J]. Science, 1989, 245:843 - 845.

[35] IRIE M. Diarylethenes for memories and switches[J]. Chem Rev, 2000, 100:1685 - 1716.

[36] KOSHIDO T, KAWAI T, YOSHINO K. Novel photomemory effects in photochromic dye-doped conducting polymer and amorphous photochromic dye layer[J]. Synth Met, 1995, 73:257 - 260.

[37] KANEUCHI Y, KAWAI T, HAMAGUCHI M, et al. Optical properties of photochromic dyes in the amorphous state[J]. Jpn J of Appl Phys, Part 1, 1997, 36:3736 - 3739.

[38] FERNANDEZ A A, LEHN J M. Optical switching and fluorescence modulation properties of photochromic metal complexes derived from dithienylethene ligands[J]. Chem-Eur J, 1999, 5:3285 - 3292.

[39] NORSTEN T B, BRANDA N R. Photoregulation of fluorescence in a porphyrinic dithienylethene photochrome[J]. J Am Chem Soc, 2001, 123:1784 - 1785.

[40] MYLES A J, BRANDA N R. Novel Photochromic Homopolymers Based on 1,2-Bis(3-thienyl)- cyclopentenes[J]. Macromolecules, 2003, 36:298 - 303.

[41] MULLER C, GIMZEWSKI J K, AVIRAM A. Electronics using hybid-molecular and mono- molecular devices[J]. Nature, 2000, 408:541 - 548.

[42] MALLOUK T, KAWAI T, YOSHINO K. Novel photomemory effects in photochromic dye-doped conducting polymer and amorphous photochromic dye layer[J]. Synth Met, 1995, 73:257 - 260.

[43] HU Y B, ZHU Y, GAO H J, et al. Photoregulation of fluorescence in a porphyrinic dithienylethene photochrome[J]. Phys Rev Lett, 2005, 95: 156803 - 156807.

[44] KAWAI T, SASAKI T, IRIE M. A photoresponsive laser dye containing photochromic dithienylethene units[J]. Chem Commun, 2001, 113:711 - 712.

[45] FUKAMINATO T, KOBATAKE S, KAWAI T, et al. Three-dimensional erasable optical memory using a photochromic diarylethene single crystal as the recording medium[J]. P Jpn Acad B-Phys, 2001, 77B:30 - 35.

[46] TOUR M P, SVEC W A, WASIELEWSKI M R. Optical control of photogenerated ion pair lifetimes: an approach to a molecular swith[J]. Science, 1996, 274:584 - 587.

[47] WOLD T, ISEDA T, IRIE M. Photochromism of triangle terthiophene derivatives as molecular re-router[J]. Chem Commun, 2004, 128:72 - 73.

[48] CHEN T, KUNITAKE T, IRIE M. Novel photochromic conducting polymer having diarylethene derivative in the main chain[J]. Chem Lett, 1999,116:905 - 906.

[49] STODDART P, HEATHY I, SIMMERER J. Electroluminescence and electron transpot in a perylene dye[J]. Appl Phys Lett, 2000, 71:1332-1334.

[50] DONHAUSER J, PFEIFFER M, WERNER A, et al. Low-voltage organic electroluminescent devices using pin structures[J]. Appl Phys Lett, 2000, 75:109 - 111.

[51] YASSAR M, KALLMANN H P, MAGNANTE P. Electroluminescence in organic crystals[J]. J Chem Phys, 2001, 38:2042 - 2049.

[52] BLUM R H. Electroluminescence from polyvinylcarbazole films Electroluminescent devices[J]. Polymer, 2005, 24:748 - 754.

[53] DINESCU L, WANG Z Y. Synthesis and photochromic properties of helically locked 1,2-dithienylethene[J]. Chem Commun, 1999, 132:2497 - 2498.

[54] HE C W, VANSLYKE S A. Organic electroluminescent diodes[J]. Appl Phys Lett, 1987, 51:913 - 915.

[55] AVIRAM S, RATNER M. Thermally irreveraible photochromic systems: A theoreti-

cal study[J]. J Org Chem, 1974, 53:6136 - 6138.

[56] METZGE M, MIYATAKE O, UCHIDA K, et al. Photochromic diarylethenes with intralocking arms[J]. J Am Chem Soc, 1997, 116: 9894 - 9900.

[57] REED P D. Redox-responsive macrocyclic receptor molecules containing transition-metal redox centers[J]. Chem Soc Rev, 1997, 18:409 - 450.

[58] ZHOU B, WANG M, WU Y, et al. Reversible near-infrared fluorescence switch by novel photochromic unsymmetrical-phthalocyanine hybrids based on bisthienylethene [J]. Chem Commun, 2002, 118:1060 - 1061.

[59] ELBING K E, WAND M D, LEMIEUX R P. Bistable ferroelectric liquid crystal photoswitch triggered by a dithienylethene dopant[J]. J Am Chem Soc, 2002, 124:7898 - 7899.

[60] DEKKER I H, DAVIDS P S, SMITH D L, et al. The schottky energy barrier dependence of charge injection in organic light-emitting diodes[J]. Appl Phys Lett, 2003, 72: 1863 - 1865.

[61] TAO M, GRABOWSKA A. How the photochromic transient is created: A theoretical approach[J]. J Chem Phys, 2000, 112:6329 - 6337.

[62] LIU Y J, HUANG M B. A theoretical study of low-lying singlet electronic states of SF2[J]. Chem Phys Lett, 2002, 360:400 - 405.

[63] APLINCOURT P, HENON E, BOHR F, et al. Theoretical study of photochemical processes involving singlet excited states of formaldehyde carbonyl oxide in the atmosphere[J]. Chem Phys, 2002, 285:221 - 231.

[64] KLENE M, ROBB M A, BLANCAFORT L, et al. A new efficient approach to the direct restricted active space self-consistent field method[J]. J Chem Phys, 2003, 119: 713 - 728.

[65] SATTELMEYER K W, STANTON J F, OLSEN J, et al. A comparison of excited state properties for iterative approximate triple linear response coupled cluster methods [J]. Chem Phys Lett, 2001, 347:499 - 504.

[66] VENTRA A B J, RETTIG W, SUDHOLT W. A comparative theoretical study on DMABN: Significance of excited state optimized geometries and direct comparison of methodologies[J]. J Phys Chem A, 2000, 106:804 - 815.

[67] GUO H, GRABOWSKA A. How the photochromic transient is created: A theoretical approach[J]. J Chem Phys, 2001, 112:6329 - 6337.

[68] XU K, WAND M D, LEMIEUX R P. Bistable ferroelectric liquid crystal photoswitch triggered by a dithienylethene dopant[J]. J Am Chem Soc, 2002, 124:7898 - 7899.

[69] EMBERLY K, SHIBATA K, KOBATAKE S, et al. Dithienylethenes with a novel

photochromic performance[J]. J Org Chem, 2002, 67:4574 - 4578.

[70] WANG C K, WANG M, WU Y, et al. Reversible near-infrared fluorescence switch by novel photochromic unsymmetrical-phthalocyanine hybrids based on bisthienylethene [J]. Chem Commun, 2002, 120:1060 - 1061.

[71] EVERS T, KUME M, IRIE M. Optical density dependence of write/read characteristics in photon-mode photochromic memory[J]. Jpn J Appl Phys, 1996, 35:4353 - 4360.

[72] LUO Y, QIU Y, WANG L D, et al. Pure red electriluminescence from a host material of binuclear gallium complex[J]. Appl Phys Lett, 2002, 81:4913 - 4915.

[73] TIVANSKI M, KALLMANN H P, MAGNANTE P. Electroluminescence in organic crystals[J]. J Chem Phys, 2006, 38:2042-2049.

[74] SEMINARI C, TOKITO S, TSUTSUI T, et al. Organic eletroluminescent devices with a three-layer structures[J]. Jpn J Appl Phys,2006, 27:713 - 715.

[75] TAYLOR J, WANG Y M, HOU X Y, et al. Interfacial electronic structures in an organic light-emitting diode[J]. Appl Phys Lett, 2005, 74:670 - 672.

[76] STOKBRO K, KALLMANN H P, MAGNANTE P. Electroluminescence in organic crystals[J]. J Chem Phys, 2005, 38:2042-2049.

[77] CHEN H, VANSLYKE S A. Organic electroluminescent diodes[J]. Appl Phys Lett, 2000, 51:913 - 915.

[78] LI R H. Electroluminescence from polyvinylcarbazole films: Electroluminescent devices [J]. Polymer, 2004, 24:748 - 754.

[79] LAKSHMI J, QIU Y, WANG L D, et al. Pure red electriluminescence from a host material of binuclear gallium complex[J]. Appl Phys Lett, 2002, 81:4913 - 4915.

[80] GALPERIN T, UCHIDA K, IRIE M. Asymmetric photocyclization of diarylethene derivatives[J]. J Am Chem Soc,2004, 119:6066 - 6071.

[81] NESS M, MOHRI M. Thermally irreversible photochromic systems: Reversible photocyclization of diarylethene derivatives[J]. J Org Chem, 2003, 53:803 - 808.

[82] WEI Y, HAYASHI K, IRIE M. Thermally irreversible photochromic systems: Reversible photocyclization of non-symmetric diarylethene derivatives[J]. B Chem Soc Jpn, 2007, 64:789 - 795.

[83] DATTA S. Elastic quantum transport calculations using auxiliary periodic boundary conditions[J]. Phys Rev B, 2005, 72: 045417 - 045419.

[84] CALZOLARI A, MARZARI N, SOUZA I, et al. Ab initio transport properties of nanostructures from maximally localized Wannier functions[J]. Phys Rev B, 2004, 69: 035108 - 035112.

[85] THYGESEN K S, BOLLINGER M V, JACOBSEN K W. Conductance calculations with a wavelet basis set[J]. Phys Rev B, 2003, 67: 115404 - 115408.

[86] FALEEV S V, LEONARD F, STEWART D A, et al. Ab initio tight-binding LMTO method for nonequilibrium electron transport in nanosystems[J]. Phys Rev B, 2005, 71: 195422 - 195426.

[87] WANG L W. Elastic quantum transport calculations using auxiliary periodic boundary conditions[J]. Phys Rev B, 2005, 72: 045417 - 045422.

[88] HAVU P, HAVU V, PUSKA M J, et al. Nonequilibrium electron trans-port in two-dimensional nanostructures modeled using Green's functions and the nite-element method[J]. Phys Rev B, 2004, 69:115325 - 115329.

[89] FUJIMOTO Y, HIROSE K. First-principles treatments of electron transport properties for nanoscale junctions[J]. Phys Rev B, 2003, 67: 195315 - 195319.

[90] LUO Y, WANG C K, FU Y. Effects of chemical and physical modications on the electronic transport properties of molecular junctions[J]. J Chem Phys, 2002, 117: 10283 - 10289.

[91] KOSOV D S. Lagrange multiplier based transport theory for quantum wires[J]. J Chem Phys, 2004, 120: 7165 - 7169.

[92] RICHARD M M. Electronic Structure: Basic Theroy and Practical Methods[M]. Cambridge: Cambridge University Press, 2004.

[93] KAXIRS E. Atomic and Electronic Structure of Solids[M]. Cambridge: Cambridge University Press, 2003.

[94] CORNIL J, KARZAZI Y, BREDAS J L. Potentials of two-photon based 3-D optical memories for high performance computing[J]. J Am Chem Soc, 2002, 124: 3516 - 3522.

[95] BORN M, HUANG K. Dynamical Theory of Crystal Lattices[M]. Oxford: Oxford University Press, 1954.

[96] HOHENBERG P, KOHN W. Inhomogeneous Electron Gas[J]. Phys Rev B, 1964, 136: 864 - 869.

[97] MAHAN G D. Many-Particle Physics[M]. 2nd ed . New York and London: Plenum Press, 1990.

[98] VANDERBILT D. Optimally smooth norm-conserving pseudopotentials[J]. Phys Rev B, 1985, 32: 8412-8419.

[99] TIVANSKI A V, HE Y, BORGUET E, et al. Electronics using hybid-molecular and mono-molecular devices[J]. J Phys Chem B, 2005, 109: 5398 - 5402.

[100] LI Z, KOSOV D S. A photoresponsive laser dye containing photochromic dithienyle-

thene units[J]. J Phys Chem B, 2006, 110: 19116 - 19122.

[101] MCWEENY R. Methods of molecular quantum mechanics[M]. London : Academic Press, 1989.

[102] TAYLOR J, GUO H, WANG J. Ab initio modeling of quantum transport properties of molecular electronic devices[J]. Phys Rev B, 2001, 63: 245407 - 245412.

[103] BRANDBYGE M, MOZOS J L, ORDEJON P, et al. Density functional method for nonequilibrium electron transport[J]. Phys Rev B, 2002, 65: 165401 - 165409.